Court-Connected Construction Mediation Practice

T0187868

The value of mediation has been widely acknowledged worldwide, as shown by the number of jurisdictions in which the courts enforce obligations on parties to negotiate and adopt mediation to settle construction disputes. This book examines the expansion and development of court-connected construction mediation provisions across a number of jurisdictions, including England and Wales, the USA, South Africa and Hong Kong. It includes contributions from academics and professionals in six different countries to produce a truly international comparative study, which is of high importance to construction managers as well as legal professionals.

Andrew Agapiou is currently Senior Lecturer in the Department of Architecture, Strathclyde University, UK. He has published refereed articles on mediation in specialist international journals such as the *International Journal of Law in the Built Environment* and *Civil Justice Quarterly*, in addition to other legal periodicals.

Deniz Artan Ilter is currently Associate Professor in the Department of Civil Engineering, Istanbul Technical University, Turkey. She graduated from the Faculty of Architecture of the Istanbul Technical University (ITU) in 2001. She specialized in contract administration in her MSc and in dispute resolution in her PhD studies, which she partly conducted at Salford University, UK.

About CIB and the CIB series

CIB, the International Council for Research and Innovation in Building and Construction, was established in 1953 to stimulate and facilitate international cooperation and information exchange between governmental research institutes in the building and construction sector, with an emphasis on those institutes engaged in technical fields of research.

CIB has since developed into a worldwide network of over 5,000 experts from about 500 member organisations active in the research community, in industry or in education, who cooperate and exchange information in over 50 CIB Commissions and Task Groups covering all fields in building and construction related research and innovation.

This series consists of a careful selection of state-of-the-art reports and conference proceedings from CIB activities.

Visit www.cibworld.nl for more information.

Recently published titles:

Construction Safety Management Systems *S. Rowlinson*
ISBN: 9780415300630. Published: 2004

Response Control and Seismic Isolation of Buildings *M. Higashino et al.*
ISBN: 9780415366232. Published: 2006

Mediation in the Construction Industry *P. Brooker et al.*
ISBN: 9780415471753. Published: 2010

Green Buildings and the Law *J. Adshead*
ISBN: 9780415559263. Published: 2011

New Perspectives on Construction in Developing Countries *G. Ofori*
ISBN: 9780415585724. Published: 2012

Contemporary Issues in Construction in Developing Countries *G. Ofori*
ISBN: 9780415585716. Published: 2012

Culture in International Construction *W. Tijhuis et al.*
ISBN: 9780415472753. Published: 2012

R&D Investment and Impact in the Global Construction Industry *K. Hampson et al.*
ISBN: 9780415859134. Published: 2014

Public Private Partnership *A. Akintoye et al.*
ISBN: 9780415728966. Published: 2015

Court-Connected Construction Mediation Practice *A. Agapiou and D. Ilter*
ISBN: 9781138810105. Published: 2016

Court-Connected Construction Mediation Practice

A comparative international review

Edited by Andrew Agapiou and Deniz Artan Ilter

Routledge
Taylor & Francis Group

LONDON AND NEW YORK

First published 2017 by Routledge

2 Park Square, Milton Park, Abingdon, Oxon, OX14 4RN
605 Third Avenue, New York, NY 10017

Routledge is an imprint of the Taylor & Francis Group, an informa business

First issued in paperback 2020

British Library Cataloguing-in-Publication Data
A catalogue record for this book is available from the British Library

Library of Congress Cataloging in Publication Data
A catalog record for this book has been requested

ISBN: 978-1-138-81010-5 (hbk)
ISBN: 978-0-367-73668-2 (pbk)

Typeset in Times New Roman
by HWA Text and Data Management, London

Contents

Figures and tables

Figures

Tables

Contributors

Andrew Agapiou is currently Senior Lecturer in the Department of Architecture, Strathclyde University, UK. He has published refereed articles on mediation in specialist international journals such as the *International Journal of Law in the Built Environment* and *Civil Justice Quarterly*, in addition to other legal periodicals. He is also part of the developing academic community researching on mediation in the UK and is a member of the UK Civil Mediation Council established by government departments, mediation providers, independent mediators and leading scholars and legal professional bodies.

Deniz Artan Ilter is currently Associate Professor in the Department of Civil Engineering, Istanbul Technical University, Turkey. She graduated from the Faculty of Architecture of the Istanbul Technical University (ITU) in 2001. She specialized in contract administration in her MSc and in dispute resolution in her PhD studies, which she partly conducted at Salford University, UK. She is the author of numerous papers published in refereed journals and conference proceedings in ADR methods, mediation, decision-making processes and AI applications in dispute resolution. Deniz has also served as a referee in her fields of expertise in several conferences and journals.

Penny Brooker is a reader in mediation at Wolverhampton University, UK. She researches in the area of commercial and construction mediation and publishes in international and national refereed journals in both the legal and built environment fields. More specifically, this research explores mediation as a dispute resolution mechanism in the construction and commercial sectors, the study of mediator techniques on the process, and the investigation into the settlement outcomes of mediation. She was joint coordinator of the International Mediation Task Group (TG 68) for the CIB (International Council for Research and Innovation in Building and Construction) from 2007 to 2010, which delivered a co-edited book on international perspectives on the development of mediation. In 2013 she completed a monograph on mediation law in England and Wales for Routledge. Recent publications review mediating in good faith, the professionalization of mediator practice, and mediator immunity.

Olive du Preez has a master's degree in quantity surveying. She is currently lecturing at the Department of Quantity Surveying and Construction Management of the University of the Free State in South Africa. Olive has also been involved in research in construction-related matters. Over the past few years she has focused her research primarily on alternate dispute resolution in the South African Construction Industry and has delivered a number of papers at various international conferences. Having initially based her research on competence levels of alternate dispute resolution facilitators in the construction industry, the findings redirected her focus toward mediation. Olive is currently studying for a PhD in conciliation, which her previous research has identified as a window for further research.

Sai On Cheung is a chartered quantity surveyor by profession. Before joining academia, he had had substantial experience in contract administration with both consultant office and construction contracting organizations. Building on these experiences, professor Cheung established the Construction Dispute Resolution Research Unit (CDRRU) (www.cdrru.org) and has developed research programs in organizational issues in construction, contract and dispute management. Professor Cheung has published widely in these areas. He received two CIOB awards for his research in construction partnering and use of information technology to minimize dispute. A collection of his research in dispute management is published as a research monograph entitled 'Construction Dispute Research'. In 2014 Professor Cheung was awarded a DSc for his research in construction dispute by the University of Salford, UK.

Laura Stipanowich is an attorney at the law firm of Smith Currie & Hancock LLP in Washington, D.C., USA, where her practice includes construction law, litigation, and alternative dispute resolution. Prior to relocating to Washington, Laura spent eight years working with a prominent law firm in Atlanta, Georgia, where she focused her practice in construction law and dispute resolution. Laura has extensive experience with the use of alternative forms of dispute resolution in the construction industry, including negotiation, early neutral evaluation, dispute review boards, mediation and arbitration. Laura has written numerous articles and regularly presents seminars throughout the United States on risk avoidance, dispute prevention and resolution, with a particular focus on issues impacting the construction industry.

Preface

The value of mediation has been widely acknowledged worldwide, as evidenced by the number of jurisdictions in which the courts enforce obligations on parties to negotiate and adopt mediation to settle construction disputes. A body of literature in the construction mediation field of course exists in many other jurisdictions, including England and Wales, Scotland, the USA, South Africa, Turkey and Australia.

This book, produced under the remit of CIB (Conseil International du Batiment (French), International Council for Research and Innovation in Building and Construction (English)) Task Group 89, builds on the work already carried out by Task Group 68, which successfully completed its initial mandate in 2010 with the publication of an edited book (Brooker and Wilkinson, 2010).This was a direct response to the expansion and development and expansion of court-connected construction mediation provisions across a number of jurisdictions, in addition to the growing body of research that examines the influences of the connection with formal civil justice systems, court-connected mediators' practices and that ways that lawyers approach mediation processes within a number of construction industry contexts. The work undertaken by Brooker and Wilkinson (2010) demonstrated the continued and dramatic expansion of mediation in construction industries across many common and civil law countries. The majority of the case studies in the review indicate that their individual construction industries have been instrumental in the development of both government and industry policies for mediation.

The work of TG 68 indicates that mediation developments are taking place extremely rapidly but not all countries are implementing mediation in the same way and the legal frameworks in some cases are substantially different. For example, some countries have approached alternative dispute resolution (ADR) and mediation by introducing mandatory rules for its use. However, mandatory legal developments are not identical and even within different countries' federal jurisdictions and courts adopt different approaches to the selection of cases for mediation (Aibinu, 2010). Both England and Hong Kong have developed similar civil procedure rules to encourage mediation and use cost penalties to encourage parties to mediate (Brooker, 2010; Sai On, 2010).

The review also reports on unique developments in some countries such as Turkey where the government is developing a legal framework to support mediation through a Mediation Act, which will regulate mediation practice and developments (Ilter and Dikbas, 2010):

- Where significant mediation developments have taken place, the steer from industry and governments has taken different forms: For example, m any common law countries have implemented court rules to encourage or use mediation and penalties for non-use.
- Some countries and specialized construction courts have their own mediation schemes or programmes to promote, govern and monitor activity.
- Other countries or federal jurisdictions have developed mandatory mediation court programmes.

The review found that these initiatives inevitably led to developing legal jurisprudence concerning the validity of contract clauses or with providing statutory interpretation of the rules requiring or governing practice. Furthermore, Brooker and Wilkinson (2010) uncovered a 'commonality of experience and approaches' (Brooker and Wilkinson, 2010, 183) where specific themes reoccurred within different jurisdictions.

Nonetheless, little was uncovered on the potential scope, purposes and practices of court-connected construction mediation and particular issues that arise within the context of formal justice systems.

This book aims to shed light on this aspect of jurisprudence worldwide. The inherent dilemma of court-connection necessarily influences the nature of mediation when it is conducted within the litigation context. Nonetheless, court-connected mediation is not restrained from delivering some degree of responsiveness, self-determination and cooperation, which are core features of the mediation process and transcend the differences between theoretical notions of mediation. These core features are delivered to disputants through opportunities to explore a range of their own interests, participate directly and cooperate. There is some empirical evidence that has demonstrated that these key opportunities are denied to many disputants within some common law jurisdictions, most notably in Australia, nonetheless, this aspect of jurisprudence remains largely unexplored within the construction context.

The bulk of academic research seeking to explore the views and experiences of key actors relative to mediation is also often concerned with the legal profession rather than clients or other players in the dispute resolution game. That this is so perhaps partly emanates from the fact that lawyers are an easier grouping for researchers and promotional bodies to access than the client base. Nonetheless, in the main, the predominant focus upon lawyers reflects the belief that they act as gatekeepers to use of mediation and hence are vital in development of the process. That this is often true is undisputable. To what extent and when lawyers are in the driving seat relative to decisions to mediate is an altogether more nuanced affair, however, that warrants further analysis and international comparison. To the best of our knowledge there is no competing book covering an international review of the legal aspects of court-connected construction mediation.

Fenn et al. edited an international review of dispute resolution and conflict management in the construction industry in 1998, which was published by E & FN Spon. This had a broad overview of dispute resolution procedures and conflict management but did not focus specifically on mediation nor did it attempt a comparative legal review of mediation. It was part of the work for a CIB commission on dispute resolution, which was the forerunner for W113. The publication of a collection of national monographs with the specific aim of describing the emerging legal challenges of construction mediation by Brooker and Wilkinson in 2010 provided a comparative commentary, for which our book seeks to build on, but with specific focus on the developing legal jurisprudence concerned with statutory interpretation of the rules requiring or governing practice within different jurisdictions and regions.

Mediation is increasingly being taught at both undergraduate and post-graduate level. We hope that the book will appeal to students of construction management, built environment and construction law, studying dispute resolution modules. We also hope that the book will have an international appeal for legal and construction practitioners, across common law jurisdictions covered in the text.

The aims of this book and its scope

This book aims to sheds light on how court-connection and lawyers' perspectives have shaped Court-Connected Construction Mediation Practice in different common law jurisdictions, specifically England and Wales, South Africa, Hong Kong and the USA.

Buth (2009) noted that with respect to the relationship between mediation and the courts, both practices and terminology vary. The term 'court-connected' mediation in this text is a 'collective' term for the many variants of mediation linked to the civil justice procedure and the court system encompassing 'court-connected', 'court-linked' or mediation of any and all matters that will of necessity be litigated.

The construction industry in England and Wales is an area where mediation has been at the forefront of development and use. Using the England and Wales (abbreviated as EW hereafter) civil procedure rules (CPR) mediation provisions as a commencement point,[1] Chapter One provides a critical assessment of the widening range of court-connected mediation possibilities available with the commercial disputes resolution sphere, within which the EW commercial mediation rules framework is considered, as a means to introduce other comparative Anglo-American jurisdiction analyses.

The EW courts, in particular the specialist Construction and Technology Court, have been instrumental in developing CPR and pre-action protocols (PAPs) which require litigants to consider using ADR before commencing litigation. Central to this strategy has been the implementation of costs penalties for unsuccessful litigants who have unreasonably refused an offer to mediate, which has led to mediation becoming an important option for construction disputes not only pre-litigation but also during the litigation phase.

In Chapter Two, Brooker shows that CPR and PAPs do not differentiate between ADR and mediation or between settlement negotiations, which blurs the line and potentially removes the imperative to mediate. CPR and the use of judicial case-management meetings are influencing litigation culture and settlement practice but it is not evident whether this is increasing the uptake of voluntary mediation other than immediately prior to commencing litigation.

Moreover, court-connection may be diminishing many of the core features and benefits of mediation, such as self-determination, speed, costs, and developing creative outcomes, as parties involve legal advisors in what is often a stage before litigation, and these lawyers introduce more evaluative practice within the process. Although the integration of mediation with court rules has raised its profile, court-connection has resulted in institutionalization. This has led to a plethora of mediation law, which is persuasive in forming party decisions about proposing or choosing to mediate, the timing and appropriateness of mediation, the costs of mediating and how the process is conducted. Moreover, as mediation becomes established practice, other legal issues come to the fore, including the enforceability of mediated agreements which encompasses the legal position on confidentiality.

The evolution of ADR in the South African construction industry has created a process which differs from standard mediation practice. Although the evolved process of mediation does not conform precisely to the pilot project of court-based mediation, it serves to relieve court congestion.

In Chapter Three, du Preez suggests that that there are mediation facilitators who will need to expand their knowledge base of conciliation which forms the basis of mediation. Mediators have concentrated on the evaluative process of mediation with limited regard for the soft skills relating to conciliation. It seems that experienced mediators support these inconsistencies by addressing the technical issues without feeling the need to consider the psychological component of conciliation. This suffices to say that mediation supported with effective conciliation may also contribute toward relieving court congestion. The practice of mediation has constantly been adjusted since its original introduction. Mediation in South Africa has slowly evolved, culminating in a hybrid form of practice. The intention was to identify a faster process for the resolution of disputes. This process, as illustrated in the Joint Building Contracts Committee (JBCC) Principal Building Agreements throughout the years, has been supported by the Construction Industry Development Board.

In Chapter Four, Cheung explores whether a direct court-connected mediation can be implemented for construction disputes in Hong Kong. Reference is made to the counterpart treatments of other jurisdictions, especially in Australia, the United Kingdom, and Singapore. Despite the advantages and benefits claimed for court-connected mediation, there are concerns that the benefits of voluntary mediation such as high settlement rates cannot be maintained with mandatory mediation. Without promising an increase in success rates, court-connected mediation can only lead to higher costs and more waiting time at the end because the disputants will need to bear both the mediation and litigation costs. In Hong Kong, contractual use of mediation is adopted for construction disputes. The use of mediation within the court system has been affected through the promulgation and application of Practice Direction 6.1(PD 6.1). Under PD 6.1, voluntary use of mediation is applied to cases under the Construction and Arbitration List with the party who unreasonably refuses or fails to mediate to the minimum participation level when there is a Mediation Notice may be subject to adverse cost orders. The enactment of Mediation Ordinance further ensures the confidential feature of mediation. Notwithstanding, there is no application of court-connected mediation for construction disputes in Hong Kong. Cheung reports that the legal profession also expressed the need to rethink the objective of having ADR for dispute resolution in Hong Kong. The overemphasis on the utilities to be derived is considered undesirable, as this would undermine the rule of law. There are a number of situations in which mediation is clearly not suitable. It is further argued that disputants insisting on litigations should not be punished for protecting their right to have disputes decided in court. With these concerns in mind, Hong Kong has not installed court-connected mediation.

The construction industry has played a major role in the development of ADR processes implemented throughout the United States of America (US). Much of the research of court-connected mediation in the US has involved formal evaluations of single programmes undertaken within individual State Court or desk studies, which describe published rules and intentions of the programmes rather than a critical analysis of the current state of court-connected programmes at the federal and state levels. It has also involved an examination of the case law and scholarly commentary.

In Chapter Five, Stipanowich provides a critique of the strengths and weaknesses of the federal and state-wide programmes, highlighting unique features, and an evaluation of their effectiveness. The traditional, often cumbersome, court system in the US did not assist practitioners in achieving these goals. Consequently, the construction industry in the US pioneered the use of mediation and modern construction participants rarely resort to traditional litigation. Stipanowich posits that in response to this sea change in dispute resolution, most US courts, both on the state and federal level, have implemented judicial and administrative procedures that incorporate mediation to supplement the traditional litigation model. Due to the distinct federal and state court systems, the mix of mediatory approaches in the US is varied and ever-changing. Stipanowich identified many differences in the mediation programmes between the various state and federal courts, but also many commonalities. For instance, a substantial number of programmes in both systems are mandatory and, as with private alternative dispute resolution, mediation is the most prevalent court-connected ADR option. The often compulsory nature of these processes has been a source of criticism and many critics question the utility of the programs. However, it seems that construction lawyers appreciate the many benefits of court-connected mediation, including judicial economy, increased efficiency, reduced cost and confidentiality. As court-connected mediation systems evolve and improve, industry practitioners also recognize them as additional tools for facilitating the economical resolution of costly litigations.

In Chapter Six, Ilter provides an overview of the development of court-connected mediation in different common law jurisdictions, uncovering the commonalities as well as the key differences in the experience and the approach in each specific country. The conclusion draws together the main points of the national case studies provided, highlighting the widening range of court-connected mediation possibilities available within the construction disputes resolution sphere. The final part of the chapter also presents a research roadmap for the further proliferation of the practice, focusing on the current barriers to the widespread use of court-connected mediation in the construction industry.

Acknowledgements

The editors would like to express our gratitude to all the contributors to the book, for their sterling efforts and hard work. This book would not have been possible without their support and encouragement. We would also like to thank the team at Taylor & Francis for finding our proposal worthwhile, and for helping us turn the manuscripts into a book. Finally, we would like to thank our families for all their patience and support.

Note

1 Civil Procedure Rules 1998 (CPR 1998) Practice Direction 'Pre Action Conduct and Protocols' (2015) [Online] Available at www.justice.gov.uk/courts/procedure-rules/civil/rules/pd_pre-action_conduct#3.1 [24 April 2016], [3], [6].

1 Court-connected mediation practice in perspective

Andrew Agapiou

Introduction

Court-connected mediation has steadily expanded its scope in many Anglo-American jurisdictions over the past 20 years.[1] Its growth has been spurred by diverse factors that include mediations' greater perceived cost-effectiveness when compared with conventional civil litigation, coupled with corresponding public, judicial and legal professional interest in the greater participant control mediation permits. This growth has been particularly evident in commercial dispute resolution, where business pragmatism is a further spur to finding mediated solutions.

Using the England and Wales (EW hereafter) civil procedure rules (CPR) mediation provisions as a commencement point,[2] this opening chapter critically assesses the widening range of court-connected mediation possibilities available within the commercial disputes resolution sphere. In the first section, key mediation theory principles are identified and explained. From here, the EW commercial mediation-rules framework is considered, as a means to introduce other comparative Anglo-American jurisdiction analyses. Cases and commentaries provide additional support for the contention that court-connected mediation boundaries are limitless, in the sense that no commercial mediation dispute is likely unsuited to a mediation attempt. The third section offers a brief prediction concerning mediation's likely future scope, where mediation will be the accepted first formal step in every commercial dispute. The conclusions section affirm the proposition that those cases where mediation fails are now regarded by many commentators as exceptions that tend to prove the rule asserted in this assessment.

Key mediation principles

Mediation may be defined as any process devised to seek settlement of a disputed issue or controversy through an independent person placed between the two contending parties to assist them. Mediation success is measured in two ways: i) where the entire dispute is settled between the parties; ii) the parties move closer to settlement ('narrowing of issues') where the foundation is established for future settlement. Mediation forms part of the now well-entrenched dispute resolution (DR) continuum. DR has evolved from its 'alternative dispute resolution' (ADR) literature references, especially when one appreciates how this formerly 'alternative' form now dominates family, tort, and commercial dispute resolution mainstreams.

The further mediation description, 'negotiating in the shadow of the law' is especially appropriate when the Heading II mandatory CPR provisions are evaluated.[3] The continuum is best understood with reference to the relative degree of formality each dispute resolution element involves. Informal, unstructured *negotiations* initiated between the two disputing parties are the crucial DR continuum commencement point.[4] Its opposite terminus is a contested *trial* proceeding, one conducted in accordance with all applicable civil justice system procedural rules.[5] Within this continuum, mediation is a middle-ground location. A mediation invariably preserves the parties' confidentiality, where each party proceeds without any commitment to reach settlement.

Mediation affords the parties opportunities for more structured information exchanges than usually occurs in privately conducted negotiation, but within a forum (and its independent third-party mediator influencing the pace and intensity of the process) with fewer procedural requirements than those governing arbitrations or trials.[6]

Mediation never forecloses the parties from abandoning its processes (consistent with fundamental disputing parties' autonomy),[7] if it is seen as likely fruitless for the respective participants, or where the parties decide to continue to a concluded settlement without mediator assistance.[8] In this important sense, mediation is an extension of its DR continuum neighbour, negotiation. Ongoing negotiation between the parties is consistent with the overarching DR commitment to promote effective, honest communication between the two parties, where an 'exchange of information, potentially leading to common understanding and joint decision-making' is a core principle.[9]

Mediation is thus concerned with effective information exchanges between the parties to permit the mediation to focus on the parties *interests*, as opposed to what legal *positions* they may have adopted prior to mediation being initiated.[10]

The first issue a mediator must isolate is the specific party interest each seeks to safeguard. Each of the mediation variants outlined here is premised on this crucial 'position versus interests' distinction.[11] As with all other dispute resolution options, mediation has distinct variants that can be further modified (or purpose-built) to suit specific dispute resolution problems.

These variants are also best understood from a DR continuum perspective. *Narrative mediation* encourages the parties to candidly 'tell their story', with the mediator using each story to construct a possible resolution pathway.[12] *Facilitative mediation* gives the mediator more scope within which to bring the parties closer to possible resolution. Confidential, 'without prejudice' joint mediation sessions are a frequently employed facilitative tool, ones designed to bring the parties' interests into closer alignment.[13]

In *evaluative mediations*, a subject matter expert mediator approaches the dispute issues from a pragmatic perspective. This mediator provides non-binding advice to the parties regarding what would likely result if their dispute proceeded to civil trial.[14] *Directive mediation* demands the greatest mediator personal engagement with the parties and the dispute issues.[15] Where the first three variants ensure the mediator remains largely above the fray, the directive mediator is expected to steer the dispute to settlement.[16] While remaining neutral, this mediator gives the parties express advice concerning what results the parties may confidently expect at trial.[17]

As suggested above, the well-recognized mediation advantages over conventional litigation or other dispute resolution methods (both through individual comparisons and where mediation is combined with arbitration, known as 'med-arb'[18]), are readily summarized. Mediation provides highly cost-effective, less formal (thus less intimidating) opportunities for the parties to meaningfully engage with each other in a neutral setting without litigation risks. Just as importantly, the fresh perspective on the dispute a mediator can provide from his or her neutral, non-binding opinion perspective can often shift the antagonistic dispute participants' attitudes.

A well-trained, issues-specific mediator can often assist the parties in getting past their emotional barriers to settlement, and encourage them to reasonably consider their settlement opportunities.[19] The disadvantages attributed to mediation are few. The most common ones cited in the mediation literature are connected to the fact that parties usually provide their opponent with insights regarding how their position will be advanced if the case is not resolved through mediation.[20]

Notwithstanding mediation's confidential, without prejudice nature, a party participating in an unsuccessful mediation knows more about their prospective litigation opponent's case than they would know otherwise.[21] Del Ceno notes that some EW civil practitioners have been resistant to mediation being imposed under CPR auspices, on the basis that such mandated DR offended a client's European Convention right of access to the courts.[22] It seems doubtful that such concerns can displace the cost-benefit and related mediation benefits cited here.[23] Specific commercial mediation concepts are now considered.

Commercial mediation has its own unique features: ones that underscore how the terms 'business', 'commercial dealings', and 'settlement' have particularly important commercial term of art characteristics that do not always apply in other DR circumstances. Bamford's 1997 article that strongly questioned commercial mediation's longer-term DR staying power seems quaint almost 20 years later, as commercial mediation has attained its niche, specialist DR status.[24]

As early as 1993, EW commercial court judges were issuing 'ADR' orders at an early stage in many large commercial disputes.[25] The 1998 CPR amendments that implemented many of Lord Woolf's 1996 'Access to Justice' report's[26] recommendations merely formalized a strongly held 'on the ground' practitioner and judicial attitude that assertive, properly structured commercial mediation tends to produce sound results.[27]

Commercial mediation has all of the advantages and disadvantages noted in the second section of this chapter, with one additional point of emphasis requiring attention.[28] The advantages are particularly apparent when compared with international commercial arbitration, the forum where most international DR is advanced.[29]

As various commentators have observed, commercial arbitration has become increasingly 'judicialized': twenty-first century commercial arbitration has evolved from a 'flexible, expedited, and less costly means of dispute settlement to a mechanism that mirrors the traditional (more costly and cumbersome) judicial process'.[30] For parties interested in commercial arbitration as a more cost-effective DR option than civil trials, there are now two uncomfortable realities. The first is that arbitration is no longer faster than litigation.[31]

The second is connected to the first, as arbitrators (unlike judges) are directly remunerated by the DR parties. As a consequence, while arbitration will often involve highly qualified expert adjudicators, it is frequently more expensive than litigation.[32] By contrast, mediation encourages a 'win-win' settlement discussions environment, where: i) mediators are less expensive than arbitrators; ii) the procedural timelines are shortened; and iii), the commercial parties' ability to potentially preserve their business relationship though a mediator's assistance is significantly enhanced.[33] The prevailing EW commercial mediation framework is now highlighted.

The EW commercial mediation framework

The CPR general provisions regarding case management apply to all EW civil proceedings, including commercial litigation.[34] The CPR establishes as its 'overriding objective' that of enabling EW courts to deal with cases justly and at proportionate cost.[35] This general CPR DR encouragement is supported by various other Rules provisions. CPR rule 1.4(1) obliges the court to further the overriding objective of enabling the court to deal with cases justly by actively managing cases.[36]

Rule 1.4(2)(e) defines 'active case management' as including the need to encourage the parties '… to use an alternative dispute resolution procedure if the court considers that appropriate and facilitates the use of such procedure'.[37] Rule 26.4(1) confirms that any civil litigation party may make a written request for the proceedings to be stayed while the parties attempt to settle their case by DR or other means.[38]

Rule 3.1(2) defines the court's general powers of management in broad, judicial discretion-rich terms.[39] Except where the CPR otherwise provides, the court may '… take any other step or make any other order for the purpose of managing the case and furthering the overriding objective'.[40] Such steps may include the court hearing an Early Neutral Evaluation (ENE) with the aim of helping the parties settle their case.[41]

The scholarly commentaries suggest that there has been some confusion in the minds of EW civil practitioners generally that the noted ENE and other DR procedures depended upon the parties' providing their consent to participate.[42] Two other reasons are cited for ENE lack of use: (i) lack of clarity regarding the basis on which courts may (with or without party consent), direct an ENE hearing be convened; (ii) under-appreciation by the judiciary and legal profession of ENE merits.[43] This reluctance has not been evident in EW commercial court practice, as there has been significant enthusiasm for ENE and other mediation-related approaches to aggressively pursue commercial claim settlements since the 1998 CPR enactments.[44]

Neville offers a brilliant assessment of commercial mediation's strengths in a specific partnership dispute context.[45] He explains how the 'reflective analysis' mediators may use in approaching a dispute between business partners provides often invaluable DR assistance, irrespective of whether the disputing parties' business is conducted within a formal partnership framework, or where the partnership is constructed as a closely held corporation.[46] Neville notes that even a seasoned mediator can find it exceptionally challenging to interact with business partners in conflict to assist them in achieving effective DR. He describes the exercise as potentially as '… daunting as Rubik's Cube and [can] make the mediator envious at the swiftness with which Alexander the Great solved the riddle of the Gordian Knot'.[47] It is the fact that EW commercial mediation growth has been steady since the 1998 Rules amendments designed to promote effective DR that confirms how readily all EW mediation stakeholders are prepared to meet the challenges Neville describes.[48]

Costs consequences

A controversial EW court-connected mediation issue is whether courts may impose cost sanctions on parties that did not meaningfully engage in mediation which the court directed under its case management mandates. This is an important issue that requires consideration of two seemingly antithetical concepts. On one hand, as outlined in the first section, *party autonomy*, mediation participation based on mutual consent is one of mediation's acknowledged strengths.[49] On the other hand, CPR rule 3 and the related Commercial Practice directions permit judges to compel the parties to participate in any mediation or other DR process the court deems appropriate.[50] These conflicting propositions have been tested in the following cases.

Halsey v. Milton Keynes[51] is the first EW appellate court consideration of whether a reluctant mediator should be punished for their deliberate decision not to fully engage with the court-directed mediation process. The interesting feature of the case which the Court of Appeal considers in this context is the fact that the reluctant party was also the ultimately successful litigator.[52] The court held that the unsuccessful party bears the onus to demonstrate why the court should sanction any departure from the general rule on costs (costs follow the event, where the successful litigant recovers their costs).[53]

Such departure would take the form of an order to deprive the successful party of some or all of their costs on the grounds that they had refused to agree to ADR.[54] The court held that, as a fundamental costs principle, a departure from the general rule was not justified unless the unsuccessful party could show the successful party had acted unreasonably in refusing to agree to DR.[55] The court (through Dyson L.J.) offers what Rix later criticizes as a less than fulsome endorsement of mediation.[56] In his two-pronged observation, Dyson L.J. first states that all EW practitioners conducting civil litigation should now routinely consider with their clients whether their disputes are amenable to a DR option. His second point is more controversial, as Dyson L.J. asserts that the court's role is to encourage DR participation (and it may encourage the parties by robust means), but the court cannot compel it.[57]

From this perspective, *Halsey* appears to endorse party autonomy, and accepted consensual mediation participation over a CPR interpretation where judges are empowered to mandate mediation at all costs.[58] The subsequent cases suggest the Court of Appeal is now prepared to sanction CPR costs orders made against parties that plainly do not agree with a case management order to proceed with mediation.

In *PGF II SA v. OMFS,* the Court of Appeal extended its *Halsey* guidelines concerning whether a refusal to engage in DR amounted to 'unreasonable conduct' that ought to attract a costs penalty.[59] The Court establishes this revised Halsey proposition as its general rule: silence in the face of an 'invitation' to participate in DR is inherently unreasonable litigation-party conduct, irrespective of whether the party advances a good reason for its DR-participation refusal.[60] The court expressly found that this commercial lease dispute was eminently suited to mediation, and that the proposed mediation had a reasonable prospect of success when it was offered by the other party.[61] In this important way, the court gives fuller meaning to what constitutes 'court-connected' mediation.

Such mediation is both DR, as expressly defined in CPR rule 3 case management, as well as mediation which a party seeks to initiate within the larger ongoing civil litigation framework.[62]

When these various EW commercial mediation points are distilled into a single impression, it is clear that the CPR framework actively encourages mediation within its broader DR parameters. The authorities suggest the EW CPR system works reasonably well, and with *Halsey* now modified by *PGF II SA* and its imposed costs consequences, meaningful mediation participation is being more than merely encouraged by EW courts. Two comparative Anglo-American jurisdiction approaches are now briefly considered.

Canadian and Australian examples contribute to a fuller understanding of how the EW court-connected mediation approaches are now reflective of clear international trends. Ontario and New South Wales are the selected jurisdictions.

Ontario

The Ontario *Rules of Civil Procedure* ('Ontario Rules')[63] provide straightforward DR direction. Under Ontario rule 24, *mandatory mediation* exists for all case-managed civil, non-family actions, including all commercial claims.[64] Mediation may only be avoided where a party secured a court order to this effect; such orders are rarely granted.[65]. A private-sector mediator is selected to conduct the proceedings, with the selection made from the Ontario Rules mediation program roster of approved mediators; the parties may agree on a non-roster mediator of their choice.[66] The Ontario procedure places some pressure on the parties, with its fixed mediation timelines. The mediator must be selected within 30 days of the first Statement of Defence being filed; the mediation *must* occur within 90 days from this filing date, unless the court determines otherwise.[67]

The mediation parties must provide the mediator and all other parties to the proceedings with their 'Statement of Issues', a document akin to an EW 'skeleton argument', that i) identifies the issues in dispute, and ii) sets out the respective parties' positions and interests.[68] This Statement must include all pleadings filed in the proceeding, as well as any other documents of central importance to the dispute.[69] The relatively strict Ontario rules have proven effective, with over 95 percent of claims eventually settling (either at mediation, or prior to trial).[70]

Ontario also endorses an equally emphatic costs regime regarding reluctant or obstructive mediation participants. In its 2010 *Keam v. Caddey* reasons,[71] the Ontario Court of Appeal ruled that where the defendant's insurer twice refused to participate in mediation, it would bear significant costs consequences. As Moroknovets observes, the court's additional $40,000 ordered in costs for failure to mediate 'makes it clear': mediation is not an option but a legitimate alternative to lengthy civil trials.[72]

New South Wales

New South Wales (NSW) establishes arguably the most comprehensive of the three jurisdictional approaches to commercial mediation cited in this analysis. In addition to its general civil procedure rules that provide for court-connected mediation on similar bases to those enacted in EW and Ontario,[73] the NSW *Commercial Arbitration Act 2010* encourages mediation as part of its larger commercial arbitration scope.[74] The NSW courts have adopted an approach to encouraging mediation that appears more aligned to the EW than Ontario experience, a point made that recognizes there is not a wide gulf in these respective approaches in any event. The NSW Supreme Court ruling in *Idoport Pty v. National Australia Bank* supports this assertion.[75]

The court found that while it has the power under the NSW rules to undertake compulsory mediation, with a corresponding parties obligation to participate 'in good faith', these rules do not mean that the parties are forced to settle.[76] The court states that if the mediation fails, the parties may continue litigation without penalty, as the NSW enactments are designed to encourage settlement rather than force it upon parties.[77] The EW, Ontario and NSW commercial mediation approached share obvious common features, most notably the respective procedural rules reinforcement of mediation's accepted importance in all modern DR processes. The likely way ahead for commercial mediation is now briefly considered.

Commercial mediation – the way ahead

Like the old expression, the 'genie is out of the bottle', there is little likelihood that the DR advances made in the past 20 years (taking a 1996 Woolf Report commencement point) will be reversed. While there may be instances where parties may have honestly held reservations concerning mediation, such as the tactical concerns outlined in previous sections, arguments underscoring collective mediation strengths are the best possible indication of future directions for court-connected mediation.

Based on how the respective EW, Ontario, and NSW jurisdictions have variously endorsed mediation, with further specific rules that assist commercial parties to more readily achieve dispute settlements, it has been predicted that it is likely that these jurisdictions will move even more assertively into mandatory mediation. It is also anticipated, as based on the authorities cited in Heading II, that each jurisdiction will enact mediation provisions that ensure mediation is the rule, with trial proceedings reserved for the truly exceptional cases. A dispute that involves a novel legal point that the parties require to be litigated, and perhaps one that should be litigated in the public interest, is the type of exceptional situation that may fall outside the ever-widening mediation ambit.

Conclusions

The cost-effective resolution of virtually any dispute is possible under court-connected mediation. This answer applies with even greater force in commercial disputes, where the parties are usually eager to ensure their business interests are not compromised by having to needlessly commit time and resources to DR processes (such as trials and their current procedure arbitration counterparts). Mediation can be tailored to suit any dispute, and the court-connected feature provides the parties (and the civil justice system administration) with a sufficiently robust process that maintains an effective balance between mediation's fundamentally consensual nature and the need to ensure disputes are brought to a timely, effective conclusion. The need to ensure that mediation is given even greater prominence is clear after careful consideration of the first section's conceptual frameworks and the second section's examples. The disputes where mediation will not succeed are very few, as long as the underlying mediation principles are matched in practice by willing parties and skilled mediators leading the process forward.

Notes

1 See (i) Anthony Connerty, 'Mediation: a scheme in operation at the Mayor's and City of London Court' (2010) 76(2) *Arbitration* 265; (ii) Anne Brady, 'Court-connected mediation' (2006) 72(2) *Arbitration* 141, 142.
2 *Civil Procedure Rules 1998* (CPR 1998) Practice Direction 'Pre-Action Conduct and Protocols' (2015) [Online] Available: www.justice.gov.uk/courts/procedure-rules/civil/rules/pd_pre-action_conduct#3.1 [24 April 2016], [3], [6].
3 Ronald Mnookin and Lorne Kornhauser 'Bargaining in the Shadow of the Law: The case of Divorce' (1979) 88 *Yale Law Journal* 950, 953.
4 Michael Supperstone 'ADR and public law' (2006) *Sum P.L.* 299, 301–302.
5 Bernard Rix, 'The interface of mediation and litigation' (2014) 80(1) *Arbitration* 21, 23.
6 Ibid. 22.
7 Rita Drummond, 'Court-Connected Mediation in England: Foundations for an Independent and Enduring Partnership (2007) [Online] Available: http://ssrn.com/abstract=962122 [24 April 2016], footnote 9.
8 Rix, (note 5), 23.
9 Drummond (note 7).
10 See e.g. D Scott DeRue, and Donald Conlon, et al., 'When Is Straightforwardness a Liability in Negotiations? The Role of Integrative Potential and Structural Power' (2009) 94(4) *Journal of Applied Psychology* 1032, 1036, 1037.
11 Ibid.
12 John Winslade and Gerald Monk, 'Narrative mediation: a new approach to conflict resolution' (Jossey-Bass, 2000), 7–8.

13 (i) Roger Fisher and William Ury, *Getting to Yes: Negotiating an agreement without giving in* (3rd edn, Random Business, 2012), xviii.
14 See John Wade, 'Evaluative and Directive Mediation: All Mediators Give Advice--Part 1 of 2' (2011) *Mediate.com* [Online] Available: www.mediate.com/articles/WadeJ3.cfm [24 April 2016].
15 Ibid.
16 Shirley Shipman. 'Compulsory mediation: the elephant in the room' (2011) 30(2) *C.J.Q.* 163, 166–169.
17 See John Wade, 'Evaluative and Directive Mediation: All Mediators Give Advice--Part 1 of 2' (2011) Mediate.com [Online] Available: www.mediate.com/articles/WadeJ3.cfm [24 April 2016].
18 Program on Negotiation 'Arbitration vs Mediation: Undecided on Your Dispute Resolution Process? Combine Mediation and Arbitration with Med-Arb' (2015) Harvard Law School [Online] Available: www.pon.harvard.edu/daily/mediation/deciding-on-arbitration-vs-mediation-try-combining-them [24 April 2016].
19 As discussed by Joshua Smilovitz, 'Emotions in Mediation' (2012) Netherlands Institute of International Relations [Online] Available: www.clingendael.nl/sites/default/files/20080100_cdsp_diplomacy [24 April 2016].
20 Shipman (note 14).
21 Drummond (note 7).
22 Julian Sidoli del Ceno, 'Compulsory mediation: civil justice, human rights and proportionality' (2014) 6(3) *I.J.L.B.E.* 286-2, citing *European Convention on Human Rights 1950*, Article 6(1).
23 Del Ceno (note 20).
24 Richard Bamforth, 'Mediation: a genuine alternative or just a passing trend?' (1997) 141(28) *S.J.* 688, 689.
25 Tamara Oyre 'Civil Procedure Rules and the use of mediation/ADR' (2004) 70(1) *Arbitration* 19.
26 Lord Harry Woolf, *Access to Justice Final Report*, (1996), [Online] Available: http://webarchive.nationalarchives.gov.uk/+/http://www.dca.gov.uk/civil/final/index.htm [24 April 2016].
27 Ibid. 21.
28 Markus Petsche, 'Mediation as the preferred method to solve international business disputes? A look into the future' (2013) 4 *I.B.L.J.* 251.
29 (i) Richard Hill, 'Adversarial Mediation' (1995) 12(4) *J. Int. Arb*. 135; (ii) Richard Hill, 'The Theoretical Basis of Mediation and Other Forms of ADR: Why They Work' (1998) 14(2) *Arb. Int*. 173.
30 Petsche, 252.
31 Ibid.
32 Ibid. 253.
33 Tony Allen, 'The place of mediation in England & Wales in 2014' (2014) 25(4) *E.B.L. Rev.* 517, 520, citing Directive 2008/52 on mediation in civil and commercial matters.
34 Justice, 'Practice Direction: Part 58 – Commercial Court' (2016) [Online] Available: www.justice.gov.uk/courts/procedure-rules/civil/rules/part58#IDAPP3HC [24 April 2016], Rule 58.3.
35 CPR 1998, r.1(1)
36 Ibid. r. 1(4) (1).
37 Ibid. r. 1(4)(20, as summarised by Rix (note 5).
38 Ibid. r. 26.4(1).
39 Ibid. r. 3.1(2).
40 Ibid. Rule 3.1(2)(m).
41 Ibid.
42 Civil Procedure News 'Early neutral evaluation' (2015) 8(Sept) *Civil Procedure News* 7, 8.

43   Ibid.
44   Khawar Qureshi, 'Money walks?' (2010) 160(7436) *N.L.J.* 1361, 1362.
45   William Neville, 'Facing the enigma: mediation and the troubled partnership' (2012) 78(1) *Arbitration* 15, 16.
46   Ibid.
47   Ibid.
48   A point taken from Peter Kelly, 'Alternative dispute resolution and the Commercial Court' (2010) 2 *A. and ADR Rev.* 92.
49   See (note 7).
50   See (note 32).
51   *Halsey v Milton Keynes General NHS Trust* [2004] EWCA Civ 576; [2004] 1 W.L.R. 3002 (CA (Civ Div)).
52   Ibid [12], [13].
53   Ibid.
54   Ibid. [8].
55   Ibid. [80].
56   Rix (note 5), 25.
57   *Halsey* (note 48), [11], [30].
58   Rix (note 5); see also Janey Draper, 'Halsey – mediation one year on' (2005) 155(7176) N.L.J. 730, 731.
59   *PGF II SA v OMFS Co 1 Ltd* [2013] EWCA Civ 1288; [2014] C.P. Rep. 6 (CA (Civ Div)).
60   Ibid. [51].
61   Ibid. [48].
62   A point taken from Patrick Taylor, 'Failing to respond to an invitation to mediate' (2014) 80(4) *Arbitration* 470, 472.
63   *Rules of Civil Procedure 1990*, as amended.
64   Ibid. Rule 24.1.
65   Ibid. see also Sue Prince, 'Mandatory mediation: The Ontario experience' (2007) 26(Jan) *C.J.Q.* 79, 82–85.
66   Ontario Rules, r. 24.1
67   See also Ontario Attorney General 'Mandatory Mediation Program' (2015) [Online] Available: www.attorneygeneral.jus.gov.on.ca/english/courts/manmed/notice.php [24 April 2016].
68   Ontario Rules, r. 24.1.
69   Ibid.
70   Shahla Ali and Felicia Lee, 'Lessons learned from a comparative examination of global civil justice reforms' (2011) 53(4) *Int. J.L.M.* 262, 274 (re Ontario Rules).
71   *Keam v. Caddey*, 2010 ONCA 565 (CA).
72   Diane Morokhovets, 'Refusal to Mediate Attracts a Remedial Penalty in Keam v. Caddey' The Court (2011) [Online] Available: www.thecourt.ca/2010/09/refusal-to-mediate-attracts-a-remedial-penalty-in-keam-v-caddey [24 April 2016].
73   *Civil Procedure Act 2005* (New South Wales), Part 4.
74   *Commercial Arbitration Act 2010* (New South Wales), s.27D.
75   *Idoport Pty Ltd v National Australia Bank Ltd* [2001] N.S.W.S.C. 427 (SC).
76   *Civil Procedure Act* (note 72), s.27.
77   *Idoport,* (note 74), [24]; see also Brenda Tronson, 'Mediation orders: do the arguments against them make sense?' (2008) 25(Jul) *C.J.Q.* 412, 414.

2 Court-connected construction mediation practice in England and Wales

Penny Brooker

Introduction

International literature describes how various jurisdictions connect mediation to courts. Such mediation ranges from judicial encouragement to participate on a voluntary basis, to court referrals mandating participation, and court orders which are sometimes called 'court-connected mediation'. (Kovach 2006, 395–398, 397; Alexander 2006). England and Wales does not have the same longevity of mediation development as, for example, Australia or the USA, both of which have mandatory court-connected programmes but nonetheless mediation is there connected to litigation through the civil procedure rules (CPR), which are applicable across the jurisdiction. CPR has stimulated extensive law relating to the 'appropriate' use of mediation and there are many legal rules which effect some of the key features of the process, such as confidentiality, the timing of mediation, and the suitability of disputes for the process (Brooker 2013; 2010a,b; 2009).

This chapter will explore how the English and Welsh jurisdiction through CPR embeds mediation into the structure of litigation by encouraging parties to take specific steps to use an alternative dispute resolution (ADR) procedure before continuing to the court. Expansion in mediation practice has been achieved by connecting the costs rules, which empower judges to penalise litigants if they unreasonably refuse to mediate in appropriate cases when one side has proposed its use.[1] By taking these measures, England and Wales, like other common law countries, have institutionalised mediation, which has led to the 'legalisation' of the process where the legal profession dominates practice both by representing the parties but also by acting as mediators (see for example, Clark 2012; Alexander 2006; Welsh 2001a,b; Brooker 2013, 33–34).

Historical context in England and Wales[2]

From the 1970s and 1980s commercial mediation was undertaken on a voluntary basis by a number of mediator devotees, many of whom had trained in the USA (Bucklow 2006). In the 1990s the courts introduced Practice Statements which required litigants, or more specifically their legal advisors, to complete check lists acknowledging that consideration had been given to ADR, and which recommended that judges issue 'ADR Orders' to stay court proceedings while the parties attempted either Early Neutral Evaluation (ENE) by a judge or dispute resolution with another neutral expert who could be a mediator[3] (Mistelis 2006, 146–147; Brooker 2013, 21–22; Brooker 2010a,b). The interest of the court was in directing disputants away from litigation to alleviate court waiting times and resources but by the mid-90s mediation became a key part of the Access to Justice Programme (Genn 2010, 33–37; 2012; Woolf 1995; 1996). Lord Woolf (1995, 1.4.1) in his review of civil justice was not only critical of the complicated litigation rules but also disparaging of the role played by the legal professions in manipulating the procedure to raise costs, create delay and thereby increase fees (see generally Mistelis 2006; Brooker 2009; 2010a,b; 2013). The Interim Report (1995, Recommendation 74) proposed a 'new ethos of cooperation' between the parties before beginning legal action and embracing ADR was to provide evidence that the litigants were cooperating with each other. The final Report (1996) made far-reaching recommendations for ADR which were encapsulated in the CPR. Central to these changes was a new legal policy requiring the courts to advance 'settlement' (Roberts 2000, 739–740; Zander 1997, 211) and pertinent to mediation is the obligation placed on the judiciary to 'actively' use case management to promote 'cooperation' which includes 'recommending' and 'facilitating' ADR in 'appropriate' cases (Brooker 2009; 2013, 26–7; 2010a, 157; Roberts 2000).[4]

CPR – cooperation through mediation

The 'strategy' for co-opting court-connected mediation are the costs rules in CPR r 44 (Brooker 2013, 38; 2010a,b; 2009). Typically the unsuccessful party has to pay the costs of the winning litigant but the court can take into account 'conduct before, as well as during, the proceedings and in particular the extent to which the parties followed the Practice Direction' which was amended in 2009 to include 'Pre-Action Conduct or any relevant pre-action protocol'[5] (Brooker 2013, 38; Brooker 2010a,b). The court will also take into consideration 'the efforts made, if any, to resolve the dispute'.[6] Therefore, if litigants do not attempt to settle or to consider mediation as a route to settlement they may be penalised in costs.

How the courts have interpreted mediation costs penalties will be considered in more detail in section 3, but an analysis of the cases indicate the creation of a body of mediation law, which has an effect on many of the core benefits that the process is said to afford the parties, such as the timing of mediation; the advantages of reduced costs for dispute resolution; the appropriate use of mediation; confidentiality and the parties' self-determination (see for example, Brooker 2013).

Evolution of court-connected mediation

Court system in England and Wales

The court system in England and Wales is adversarial in nature. It is based on a tiered structure with the Supreme Court at the head, cascading down to Court of Appeal (CA) and then the High Court which is subdivided into the Chancery, Family and the Queen's Bench Divisions (Brooker 2010a, 155–156; HMCS 2009). Construction-related cases under £15,000 (there are proposals that this should be raised to £100,000: Ministry of Justice, 2012) are usually heard in the Crown Court but if particularly complex they can be transferred to the specialist Technology and Construction Court (TCC) which is in the Queen's Bench Division[7] (Brooker 2010a, 154–5). Since 2012, cases in the London TCC under £250,000 are automatically transferred to Central London County Court or to another TCC at a different centre because judges were experiencing an escalation of 'about 75 per cent' of trials due to the court's efficiencies.[8]

Settlement and pre-action protocols connect mediation to the courts

A significant strand of Lord Woolf's plans for ADR was that courts should design Pre-action Protocols (PAPs) to deal with specific areas of litigation with the objective of emphasising the prompt settlement of disputes before legal action[9] (Woolf 1996, Chapter 9; Brooker 2013, 27–28; Gerber and Mailman 2005). One of the first cases in the TCC which illustrates how ADR is connected to litigation is *Paul Thomas Construction Ltd v. Hyland* (2000),[10] when a builder offered to use ADR with the homeowners but the court found he had made 'unreasonable' conditions over the payment of the neutral and indemnity costs were awarded for contravening court protocols by beginning proceedings without considering other methods of settlement (Brooker 2013, 39; Brooker 2010a,b, Gerber and Mailman 2005, 248–249).

Technology and construction pre-action protocol

The latest amended PAP for Construction and Engineering Disputes was published in 2014 and is applicable to 'all construction and engineering disputes (including professional negligence claims against architects, engineers and quantity surveyors) (hereafter known as TCC Protocol).[11] The TCC Protocol is the only one which requires the litigants to participate in a Pre-action Meeting before issuing proceedings where they have to discuss the issues between them and consider whether there are other methods of settling all or parts of the dispute[12] (TCC Court Guide 2014, 2.4.3; Brooker 2010a, 157–158; 2013, 27–28). Section 5.4 obliges litigants to deliberate on whether ADR would be more appropriate than court and if so to determine what type of dispute resolution process to use but also expressly states that 'no party' should be compelled to go to mediation or any other ADR procedure.

Court management conference

Following the pre-action meeting, if no resolution has been attained, proceedings are commenced in the TCC after which the litigants are contacted within 14 days to set up the first Court Management Conference (CMC).[13] There are a number of objectives for this meeting, such as issuing appropriate orders or evaluating the need for expert witnesses but it also provides the opportunity for the judge to explore the 'outcome of the protocol process' and to question whether 'further ADR' is possible.[14]

The TCC Court Guide provides information on ADR and states that at the CMC the parties will have to 'address' the court on the 'efficacy of ADR'. It is at this stage that an ADR order may be issued with a 'Stay of Proceeding'.[15] The court may also at any period up to the trial, either on its own 'initiative' or the request of a party, invite the litigants to reconsider ADR or issue a 'short stay' for ADR to take place.[16] The guide defines ADR as 'any process through which the parties attempt to resolve their dispute',[17] which can be 'inter-party negotiation' or mediation but also 'neutral evaluation by a judge'.[18]

Many TCC judges are renowned for their support of ADR and mediation is often recommended to the parties as a course of action (Brooker 2009, 89–90; 2010a, 161; 2010b; 2013, 142–6; Brooker and Lavers 2002). For example, in *Brookfield Construction (UK) Ltd v. Mott MacDonald Ltd* [2010][19] after two case management meetings with the parties Coulson J. issued a vigorous warning to end their 'uncooperative' approach and to consider some type of ADR because the size of the dispute (combined claims of over £75 million) involving 'finance and documents' made it 'ideally suitable' for mediation and moreover both litigants were forewarned that a failure to mediate by one party would be considered in costs (Brooker 2013, 146).

TCC Court Settlement Process (Service)

Since 2005 the TCC has run a Court Settlement Service (CSS) where the court will provide at the parties' request a non-binding evaluation of the case by the judge or other neutral, or mediation with a trained judicial mediator[20] (Gould et al. 2010, 3). The CSS is commenced after a Court Settlement Process Order is issued:[21] Appendix G of the Court Guide provides that the judge can 'conduct' the Court Settlement Process in 'any way he deems fit' but will take into consideration the 'parties' views' but if no settlement is reached the parties' can request from the judge an opinion on the case, its 'likely outcome' at trial or 'what is appropriate for settlement'.[22] Evaluation of this nature is likely to be highly instrumental in reaching settlement but it has not been possible to find any data from the TCC on this process. The TCC Annual Report (2014) did not provide information on how many judicial mediations have taken place but a mediation survey in the TCC (Hereinafter TCC Survey) undertaken from 2006–2008 found only five respondents had used a judge as a mediator from the CSS indicating that it is not a common choice (Gould et al. 2009, 15).

A number of surveys shed some light on construction mediation (Brooker and Lavers 2005a,b; Agapiou 2015; Gould et al. 2009, 2010). Brooker and Lavers' study (2005b, 14–15) found that the majority of reported mediations were 'initiated' by the parties (76 per cent) and only 12 per cent were activated as a result of an 'indication from the court' with a further 10 per cent through an

ADR Order. The TCC Study found that settlement during litigation is mainly due to 'conventional negotiation' (66 per cent) followed by mediation (26 per cent) which was mostly undertaken in the early stages of proceedings (Gould et al. 2010, 49). The research found there were three stages when mediation was most likely to occur at the parties' own instigation: First during 'exchange of pleading', second 'during or after disclosure' and third just before trial (Gould et al. 2010, 63). When mediation ensued due to court encouragement, the data suggests this might follow after pleadings because of the CMC or just before trial because of 'a Pre-Trial Conference' (Gould et al. 2009, 18). The overall conclusion of the TCC Survey was that CPR and the Pre-action Protocols were working to encourage mediation use (Gould et al. 2010, 63).

ADR Orders – stay in proceeding to encourage and facilitate mediation

One way that courts encourage mediation is through issuing ADR Orders under CPR 26(4), which is usually activated through the 'direction questionnaire' or by 'written request' from the parties or during the CMC (Brooker 2013, 67–73). A typical stay is for one month but CPR 26.4(2A) stipulates it can be for a period that is 'appropriate' and can be extended (s26.4(3)). Once an ADR Order has been issued, the 'claimant' is required to inform the court if the dispute settled (26.4(4)). Early case law indicated inconsistency on whether the court could issue an ADR Order without the consent of both parties but the CA in *Halsey*[23] [2004, 10] held that litigants cannot be forced to attend mediation as this would undermine its 'voluntary nature', which is the basis of its 'effectiveness' (Brooker 2009, 2010a,b, 2013; Shipman 2006; Brunsdon-Tully 2009). *Halsey*[24] held that ordering mediation without consent would contravene a person's right of access to court under the Human Rights Act[25] but this has been challenged in judicial speeches (Clark 2008; Phillips 2008; See Brooker 2010a, 170; 2013, 71). Recent cases call for stronger court powers to compel attendance, which appears to be based on the view that once mediation is experienced then it can often lead to 'unexpected' success.[26]

Sir Alan Ward opined in *Wright v. Michael Wright* [2013] that it would be better if the courts are able 'to shift intransigent parties off the trial track onto the parallel track of mediation' with a court ADR Order even if they did not consent, because it would not 'really' impede their 'right of access to the court' for long.[27]

An analysis of the case makes it difficult to understand on what basis mediation would work as Sir Ward observed that you may be able to force the 'mule' to water but not to drink: "I suppose you can make it run around the litigation course so vigorously that in a muck sweat it will find the mediation trough more friendly and desirable. But none of that provides the real answer".[28] The Court of Appeal recommended that the parties consider mediation before retrial but one might query if there is limited finance to appoint lawyers to assist with litigation, as the case involved parties representing themselves, there may also be a lack of resources to pay the average daily fee for experienced mediators. CEDR (Centre for Effective Dispute Resolution) (Mediators' Audit, 2014) indicate experienced mediator fees are now £3,820 (down by 10 per cent on 2012) or £1,500 for those with less experience (also slightly down on 2012). Parties who have invested in litigation may feel that they have to continue in the expectation or hope of winning and thereby recovering costs from the unsuccessful litigant. The decision is likely to be influenced by the outlay needed for mediation, particularly when legal representation is used, which may not be recoverable as the parties usually share the costs of mediating.

Recovering the costs of mediation

The parties are able to recover mediation costs in a number of circumstances both 'post-action' and 'pre-action' (Brooker 2013, 156–16; White Book 2015, s14.5). First, judges have discretion in costs which include those that are 'incidental' to litigation which may be mediation when an ADR Order has been made.[29] Second, CPR r 5.6 Part 23 permits a claim for costs 'in connection with negotiation with a view to settlement' which can apply to mediation because the process is 'analogous' to negotiating.[30] Costs can also be claimed for 'pre-action mediation' when they are 'incurred in compliance with a 'Pre-Action Protocol'[31] but the award may depend on how long before litigation the parties mediated[32] and their understanding that they are participating in the process as 'part of the pre-action protocol'[33] (see Brooker 2013, 16).

Brooker observes that the 'timing' of mediation is likely to be influenced by the possibility of recovering the costs (Brooker 2010b; 2013, 161; Brooker 2010a, 154) as parties may be discouraged from mediating until they commence litigation (Brunsden-Tully 2009). This creates a 'symbiotic relationship' between litigation and mediation which has been observed to lead to the parties engaging in a 'tactical interplay' with the two processes (Brooker 2013, 161, 183; Brooker 2010a,b; 1999). This issue is considered more fully below.

Numbers of construction mediations

It is virtually impossible to estimate how many construction mediations occur each year as there is no central system for collating this data (Brooker 2010a; 163–164; Gould et al. 2010). CEDR (Mediators' Audit 2014, 3) reported a 9 per cent increase in 2013, bringing the total annual mediations to approximately 9,500. CEDR (Statistics (2003) indicated that construction disputes comprised 6–8 per cent of their mediations therefore it is estimated that 570–760 construction mediations take place annually by mediators completing the survey (Brooker 2010a, 163–164).

Court statistics do not indicate how many ADR Orders are issued each year but the TCC Survey (2010, 50) found 22 per cent of mediations were prompted by court orders or 'indicated' by the judge. The annual TCC Report (2014, 5) states that the court is often reluctant to issue an ADR stay if this jeopardises the trial date which suggests that settlement rather than any other benefits afforded by mediation is the driving policy behind encouraging ADR (Genn 2010; Roberts 2000).

ADR is thus connected to litigation by bringing it to the attention of litigants during the pre-action process, the CMC, and the run-up to trial. Allowing parties to recover pre-action or post-action costs for mediating further promotes usage. The majority of court rules do not differentiate between alternative procedures, which has blurred the edges between settlement negotiation, mediation and dispute resolution involving evaluation. For example, the TCC Court Guide (2014) states that ADR embraces any procedure 'where the parties attempt to settle' including 'inter-party negotiations', this may include Early Neutral Evaluation with a judge or other neutral evaluating the case for the parties.[34] The Court Guide asserts that the TCC PAP is a form of ADR,[35] which, while encouraging the parties to cooperate, may also diminish the utilisation of mediation (Brooker 2013, 131–132). There are also cases where the losing party has not been penalised for refusing mediation because reasonable offers had been made to settle as well as a request for 'round table discussions'[36] or when an 'honest broker' was appointed,[37] which led Brooker and Lavers to question whether a broad classification of ADR would lead the way to accepting 'lawyer-negotiations' as an adequate substitute for mediation (Brooker and Lavers 180–182; Brooker 2013, 125; 2010a,b).

In 2015, the White Book, which is the leading manual on the litigation procedure, suggests that in the future the question of adverse costs will not involve refusing to mediate but will be incurred for not following 'case management directions' (14.11): "The expectation seems to be that active case management will render mediation part of the normal pre-trial case management process, so it should not be necessary for the courts to make frequent use of cost sanctions". Such an approach may positively influence litigation culture and settlement practices but to what extent it will increase voluntary mediation before legal action is difficult to assess.

Certainly, parties and their legal representatives know that arriving at the CMC without deliberating on ADR may lead to court directions to consider mediation or even to a short stay in proceedings, but when mediation is connected to the courts in this way it affects the practice of mediation, the model of mediation, its timing and potentially settlement outcomes. This is explored in the following section.

Core features of court-connected mediation

The core features of mediation have been summarised as 'self-determination, autonomy, empowerment, transformation, and efficiency' (Nolan-Haley 1996, 54), which have to be encapsulated by some to provide the benefits of: 'consensus; continuity, confidentiality and control' (O'Connor 1992, 108; Dixon and Carroll 1990).

Self-determination, autonomy, empowerment and transformation

Many writers consider the 'core value' of mediation to be the parties' self-determination (Alfini 2008, 830–31; Imperati 2007, 647; Nolan-Haley 1996, 55; Welsh 2001a, 8). Mediation is said to 'provide a forum' where the parties are pivotal to the resolution of their own dispute from shaping the 'structure 'of the process (Nolan-Haley 1996, 55), framing their own 'substantive norms for decision making' and 'creating' their own settlement outcomes or even leaving the process without any agreement; and the mediator's role is one of supporting the parties' self-determination (Welsh 2001a, 4). Autonomy is said to 'underlie' the principle of self-determination but only when the parties are presented with the choice of whether to use 'alternatives' or settle their dispute (Nolan-Haley 1996, 90 citing Matz 1994).

Party autonomy and self-determination may have been the original concepts behind the 'Modern Mediation Movement' (Alexander 2006, 1) but there is little evidence that either were key factors in formulating CPR, nor do they appear fundamental to court-connected mediation as it has evolved in the English jurisdiction because, although the parties retain choice over whether to mediate, this control is fettered by court rules and judicial encouragement which restrain genuine autonomy. While CPR encourages the parties to explore settlement opportunities and alternatives to litigation, the use of costs penalties undermines both self-determination and autonomy as parties may only elect to mediate because of an apprehension that a refusal will signify unreasonable behaviour (Brooker 2013; 2010a,b, 2009; Brunsdon-Tully 2009, 232; Shipman 2006, 192–193).

Impact of mediators' orientation

For some theoreticians, a mediator's role is less about 'settlement or problem solving' but about 'empowering the parties' (Welsh 2001b, 90) through mediation's 'transformative effect', which supports the parties' "analysis and decision making (the "empowerment" effect)' and (2) enables them to 'appreciate perspectives different from their own (the "recognition" effect)' (Noce et al. 2002, 50; Folger 2002, 729–730; Baruch-Bush and Folger 1994).

Research suggests that there are few construction mediators in England and Wales who practice purely transformative mediation, and a review of training and mediator profiles indicates that it is likely that most use a mix of facilitative and evaluative techniques (Brooker 2007; Brooker 2011). As mediation is more closely connected to litigation, research in England and other parts of the UK indicates that evaluation is often an integral part of mediator practice and construction mediation has not been immune to this development (Brooker 2007; Clark and Agapiou 2011). Evaluative mediation has the effect of reducing self-determination as mediators take control of the process by providing settlement options, evaluating the 'strengths and weaknesses' of the participants' cases or 'predicting' likely outcomes at court (Kovach and Love 1996, 1998; Welch 2001a,b). Brooker raises concerns that parties in EW engage in mediation without sufficient 'informed consent' about such issues as confidentiality and self-determination which generates further doubt on party autonomy (Brooker 2011; 2013, 244–255).

The practice of mediation is not regulated in England and Wales, although many ADR organisations provide mediator qualifications through successful course completion or have their programmes accredited by Civil Mediation Council (Brooker 2011, 29; 2013, 250–255; Boon et al. 2007; Gould et al. 2010, 11–13). There is no requirement that mediators must embrace any specific approach to mediation, although members must adopt as a minimum the EU Code of Conduct but this also does not stipulate any mode of practice, nor does it identify self-determination as a core value but it does state that mediators 'must ensure that all parties have adequate opportunities to be involved in the process' (Brooker 2011, 29).[38] Currently, it is possible to say that England and Wales's mediators are trained in facilitative mediation but not exclusively, and they are not barred from employing evaluative interventions (Brooker 2011). However, further research is needed to establish the paradigm of construction mediator practice and its effect on party participation and outcomes.

Efficiencies

The key promotional features of mediation are said to be the 'efficiencies' that can be achieved in terms of a speedier way to reach settlement while at the same time delivering on savings for dispute resolution and management expenditure compared to the high costs required in pursuing a case in litigation (Brooker 2013, 9–11; Genn 2010; Gould et al. 2010).

Settlement outcomes

A major marketing device has been the high settlement rates that are said to occur, which thereby avoid the costs involved in pursuing litigation. In the early days of mediation, reports of settlement rates of nearly 90 per cent were not uncommon, although it is not easy to establish the accuracies of such claims because of a lack of data and recording methods (Brooker and Lavers 2005a 184; 2005b; Gould et al. 2010; Genn 2010, 2012; Brunsdon-Tully 2009). Mediation with its apparent high settlement rates became a key consideration in relocating disputes away from the court to the private sector (Roberts 2000; Brooker 2013, 26).

Finding accurate settlement rates for mediation is also hindered because records sometimes include agreements reached after the process has come to end and the time frame (days, weeks or months) is often not available, making it impossible to state how long the process stimulates later resolution of the dispute. Mediation is also reported to facilitate 'partial' settlement by narrowing some of the issues that will be then taken forward to trial, which also obscures settlement rates (Brooker and Lavers 2005a, 187–190; Gould et al. 2010, 25). Mediators in CEDR's Audit (2014, 8) reported a 75 per cent settlement rate with a further 11 per cent settling 'shortly thereafter', which shows little variance from the previous reports but data is not given for different areas of mediation. Brooker and Lavers (2005a, 188) found that construction disputes settled significantly less often than commercial ones (70 per cent compared to 81 per cent) but construction mediations reached partial settlement more often. The findings revealed that disputes in the construction industry were often technically complex, involved multiple parties and several issues, making them amenable to hiving off parts of the dispute for resolution (Brooker and Lavers 2005a, 188). Settlement rates in the CA mediation scheme are reported to be 68 per cent (Judiciary Website, 2015) but figures for different fields of dispute are not provided. A survey of the CA scheme found there was a 45 per cent rate between 1997–1999 which led Genn (2002, 98) to conclude that after one party had already won at court there are few 'expectations' from mediation and reduced 'scope' for compromise.

The TCC Survey found that 35 per cent of respondents reached settlement through mediation once litigation proceedings had started, in comparison to 65 per cent through 'conventional negotiation' and although 20 per cent of those who experienced a failed mediation attempted mediation just before the hearing, only 9 per cent reported that some issues reached settlement (Gould et al. 2010, 49, 54).

Costs

The majority of cases never proceed to trial, therefore Genn (2010, 112) observes that it is difficult to assess what costs have been saved by mediating compared to litigating. The TCC Survey did not provide comparisons of the financial savings between mediation and using 'conventional negotiation' but mediation is likely to be more expensive because the cost of the mediator, the venue and legal representation have to be factored in. Nevertheless, Gould et al. (2009, 66) ascertained that court-connected mediation made savings of £25,000 or lower for 15 per cent of the respondents; £25,000 to under £300,000 for 76 per cent, and 9 per cent saved over £300,000. It is probably not surprising that 25 per cent of those who continued to trial because of non-settlement considered mediation to be a 'waste of time', 25 per cent felt that it was a 'waste of money' and 10 per cent agreed that it caused a 'delay in the litigation timetable' (Gould et al. 2010, 60).

Creative outcomes and continuing relationships

Mediation is said to provide the opportunity for a better outcome for disputants because it permits the parties to devise a 'creative outcome' which does not have to be based on the legal position and can involve either 'new business or commercial arrangements' or 'continuing relationships' (Brooker and Lavers 2005a,b; Alexander 2006 11; O'Connor 1992; Stipanowich 2001, 849).

The effect of mediating is said to preserve or help to repair the relationship which adversarial litigation does not achieve (Goldberg et al. 1985).

Little empirical research has explored the outcomes of construction mediation in England and Wales, although Brooker and Lavers (2005a, 32–33) found that the majority of disputes (77 per cent) resulted in 'financial settlements': 14.7 per cent in a mix of financial and creative outcomes and 3 per cent achieved only a creative outcome. In follow-up interviews, nearly all of the lawyers reported that they had not been involved in mediation which resulted in a continuing relationship and a number felt that this was unlikely, particularly when the parties had already embarked on litigation.

The effect of CPR, the protocols and costs, may be leading to more mediations occurring after the parties begin litigation rather than at the onset of the dispute (Brooker 2009; 2010a,b; Gould et al. 2010). Agapiou (2015, 241) found that there may be difficulties in organising mediation when multiple parties and their legal representatives have to be accommodated but only 4 per cent of lawyers surveyed said that the lack of availability of 'good' construction mediators 'often' effects whether they recommend mediation to their clients and 64 per cent said this 'rarely or never' influenced their recommendations.

Positive and negative benefits

The TCC Survey found that even when settlement was not achieved, 10 per cent of the respondents considered mediation had enabled them to 'narrow' the issues and a quarter believed the process enabled a 'greater understanding' of the dispute (Gould et al. 2010, 60). This confirms Brooker and Lavers' (2005a, 119; Brooker 2007, 149) research which found a range of 'ancillary benefits' including reducing issues in the dispute or enabling a better understanding of the 'strengths and weaknesses' of the case, but the study also highlighted other more cynical gains such as: 'testing evidence' or 'witnesses' before going to trial, delivering 'messages', 'eyeballing the other side', or getting a 'feel of the financial muscle' of their opponents or even utilising mediation to create delay which has been judicially commented on in construction cases (see for example, Brooker 2010a 148, Brooker 2009; Brooker 2013 132, 148–154; Tully-Brunsdon 2009; Shipman 2006).[39] Half of the construction lawyers in Brooker and Lavers' study who had experienced an unsuccessful mediation stated that the tactical use of the process had contributed to the failure, although only six respondents said that it had been used for delay, but an analysis of comments exposed concerns about delay and the wasted costs when settlement was not achieved (Brooker and Lavers 2005a,b).

Despite accusations that some parties and their legal advisors use mediation strategically there is evidence that participants in construction mediation are satisfied with their experience. In Brooker and Lavers' survey (2005a, 202–205) 72 per cent of construction lawyers declared satisfaction with the mediation process, 79 per cent with the outcome, 84 per cent with the speed, 66 per cent with the costs, and 69 per cent with the mediator.

In Agapiou's research (2015, 238) 64 per cent considered the process to be 'sometimes' 'effective for resolving disputes'; 20 per cent 'often' and 8 per cent 'always' but a negligible 2 per cent reported it was 'never effective'. Moreover, 78 per cent who had 'negative experiences' mediating said that this had 'little' or 'no effect' on whether they recommended the process to their clients (Agapiou 2015, 241).

Commentators suggest the benefits in reducing the pressure on court resources is the dominant incentive for the implementation of legal policies for court-connected mediation rather than the advantages that accrue to parties (Genn 2010; 2012; Brunsdon-Tully 2009; Hensler 2001, 2003). With the exception of the Small Claims Mediation Scheme, there is no central body responsible for bringing together data on mediator activity or mediation outcomes (Brooker 2010a,b; 2013; Gould et al. 2010) The collection and dissemination of such information would be useful for court-connected mediation policy in the construction field.

Evolving mediation case law

Court-connected mediation has had a significant impact on the development of mediation law which has grown substantially since the implementation of CPR. Perhaps one of the most influential CA decisions on mediation practice is *Halsey* which explicates the guidelines on when the court will sanction parties for unreasonably rejecting an offer to mediate. A detailed analysis of the *Halsey* criteria indicates that court decisions are now 'normalising practice' as the guidelines are used to 'mount or lead arguments against adverse costs'[40] (Brooker 2013, 166; Brooker 2010a, 151). *Halsey* provides 'benchmarks' for using mediation by providing six 'non-exhaustive' circumstances which the courts will take into consideration when considering the 'suitability' of mediating (Brooker 2009; 2010a,b; 2013; Brunsdon-Tully 2009; Shipman 2006).

1 NATURE

Although Halsey and subsequent cases suggest that most disputes will be suitable for mediation, specific factors, or the 'nature' of the dispute may indicate that the party has not unreasonably refused ADR: For example, when a litigant or the law requires a legally binding decision or an 'injunction' or when there is the existence of fraud, this may be evidence that the party has been reasonable in proceeding to litigation.[41] Construction disputes of 'small value' are recognised as being particularly suitable for mediation42 or where there are numerous 'documents' or when the case has a number of issues parts of which might be resolved43 (see for example, Brooker 2013, 144).

2 MERIT

The criterion that is perhaps most difficult to evaluate is when a party is claiming that the 'merit' of their dispute is such that they will succeed at court, which *Halsey* suggests may place large businesses under pressure from 'costs threats' or 'speculative claims'[44] (Brooker 2009, 87; Brooker 2013 at 137; Barratt 2008).[45] The risk of this factor is the danger of wrongly assessing the strength of the case (Shipman 2006, 204), particularly in 'borderline' situations when the court may find a refusal unreasonable.[46]

3 USING OTHER ADR METHODS

Parties may also succeed in proving that they have not been unreasonable if they are able to show that they have already attempted 'other' dispute resolution alternatives or made 'settlement offers'.[47] Brooker (2013, 131) explores how this reason may have done little to promote mediation as it 'validates settlement negotiation', which is often undertaken under advisement of legal professionals who may use 'adversarial' rather than 'principled negotiation' techniques.

4 COSTS

Halsey (2004, 21) considered that 'ADR costs' may provide a valid excuse, particularly when the dispute is not 'disproportionately high' but it was noted that the outlay for mediation is not negligible. Suter (2015) observes a recent trend in the case law which discounts the relevance of this criterion. In *Northrop Grumman Mission Systems Europe Ltd v. BAE Systems Ltd* [2014] the TCC estimated that the expense of mediation would be £40,000, which was thought to be good value for a dispute involving £3 million and when costs had already reached £500,000.[48] In *Garritt-Critchley and Others v. Ronnan and Solarpower PV Ltd* [2014][49] a much smaller claim of £208,000 was 'revised' to £10,000 close to trial, which the defendants believed would be about the same amount as a day at mediation, but the court rejected the argument that costs were disproportionate because the calculation should have been between mediation and a day at trial, and it was suggested that litigants would be better to mediate small financial disputes (Sidoli del Ceno 2015; Suter 2015). Suter (2015, 10) observes that 'in all but the most modest of claims, cost will not be a realistic deterrent to mediation'.

Mediators' fees have risen dramatically since the early days of mediation and observers note that when court-connected the parties are more likely to bring lawyers, particularly when the amount in dispute is high, which not only increases costs but also reduces self-determination as legal representatives begin to 'dominate' the process (Brooker 2010a,b, 148; Shipman 2006, 147–8; Genn 1998, 2002; Brooker and Lavers 2005a,b; Clark 2012, Kovach and Love 1996, 2008; Riskin 1996; Welsh 2001a,b).

5 DELAY

Halsey[50] stipulates that if using mediation would create a 'delay' in reaching trial then a party may be justified in refusing the offer. An analysis of later cases illustrates that determining the best time for mediation is not a precise art nor is it possible to anticipate when the courts will accept that it is reasonable to defer mediating[51] (Brooker 2009; 2010a,b; 2013; Shipman 2006; Tully-Brunsdon 2009). Cases indicate that parties may be justified in declining mediation 'very late' in the litigation process[52] or when parties have not provided sufficient 'particulars of claim' or 'reasonably requested information'[53] (Brooker 2013, 153–154). Furthermore, the TCC recognise that sometimes parties find themselves in 'costs difficulties' and make 'tactical mediation offers' to avoid later costs penalties[54] (Brooker 2009, 93; 2010, 168; 2010a, 148; Brooker 2013 132, 148–154; Brunsdon-Tully 2009; Shipman 2006). Brooker (2013, 148–149) states that litigation procedures require the parties to engage in a 'risky balancing act' on when they should mediate: 'The pre-action protocols drive the litigation process through the exchange of pleadings, discovery of evidence and witness statements and entangles the timing of mediation within this framework'.

6 'REASONABLE PROSPECT OF SUCCESS.'

A factor which is equally difficult to appraise is the party's belief whether mediation has 'any reasonable prospect of success, particularly when it is based on the conviction that the other party's' 'obdurate attitude' will negate any possibility of success[55] (Brooker 2009; Brooker 2010a, 167; Brooker 2013; Shipman 2006; Tully-Brunsdon 2009). Brooker observes that a party's confidence in their case or a desire to have a court decision do not typify 'unreasonable conduct' but when these views are unreasonably held they are 'unjustified' (Brooker 2010b, 151; 2013, 133). *Halsey* states that 'borderline' cases are likely to be suitable, which means that when parties are not certain of litigation success then the court are likely to regard a decision not to mediate as unreasonable, particularly as one of the benefits of process is enabling to parties to review the 'strengths and weaknesses' of their dispute (Brooker 2009, 8).[56]

Extension to Halsey's guidelines

The CA in *PGF 11 SA v. OMFS Company Ltd* [2013][57] extended the *Halsey* criteria to hold that it is unreasonable not to reply to mediation offers or give a 'flat rejection' (Meggitt 2014; Suter 2015; Sidoli del Ceno 2012; Allen 2012). The court took note of a number of factors including the 'suitability' of most disputes for mediation, the high settlement rates reported by CEDR, Lord Jackson's Costs Report (2009) on the value of mediating because the protocols consideration of mediation and lawyers' have a duty to provide information, all of which indicate that the parties should not just 'ignore' ADR offers[58]. Finally, the court referred to the ADR Handbook, initiated by Lord Jackson's Report which gives advice to parties on what they must do when an ADR offer is made (Blake et al. 2013):[59]

a) Not ignoring an offer to engage in ADR;

b) Responding promptly in writing, giving clear and full reasons why ADR is not appropriate at the stage, based if possible on the Halsey guidelines;

c) Raising with the opposing party any shortage of information or evidence believed to be an obstacle to successful ADR, together with consideration of how that shortage might be overcome;

d) Not closing off ADR of any kind, and for all time, in case some other method than that proposed, or ADR at some later date, might prove to be worth pursuing.

The Court of Appeal condensed this advice to requiring litigants to participate in 'constructive engagement in ADR rather than flat rejection, or silence'.[60] Authoritative court statements will inevitably filter down and normalise legal practice. However, this extension is only a 'general rule' in order to cover 'rare cases where ADR is so obviously inappropriate' and possible 'mistakes in the office' such as not knowing that an offer existed.[61] The reasoning behind the decision is the problem presented to the court in determining 'belated' explanations for not using mediation and the desire to send a strong message to future litigants that they 'must engage in a serious invitation to participate in ADR'.[62]

Meggitt (2014) observes that recent case law is making mediation all but mandatory because many of the objections, which previously may have justified a refusal under *Halsey*, have not been accepted either with limited court analysis or with robust declarations that mediators can resolve even the most difficult cases (Suter 2015, 10; *Halsey* 2004, 27, Rix 2015, 24–25).

For example, an examination of *Northrop Grumman Mission Systems Europe Ltd v. BAE Systems* (2014)[63] reveals that although the court recognised the defendant had a strong case and that parties should be permitted to protect themselves from nuisance claims, nevertheless it was stated that mediation could help even when there was 'no merit' in a case because a 'mediator could bring a new independent perspective' or bring about a creative outcome by using 'evaluative' and 'facilitative' techniques (Meggitt 2014; Sidoli del Ceno 2015).[64] The court dismissed the argument that the refusal was justified because one party wanted a legal decision on a contractual point as a 'continuing relationship' did not rely on this and a mediator may have helped them to resolve the issue.[65]

CPR and the protocols have raised the profile of mediation but the cases show that there is now an 'interplay' between litigation and mediation strategy (Brooker 2010b, 149; 2013, 262). In some areas the law is failing to establish clear guidelines, particularly when the case involves delaying litigation, which has the potential to lead to unfairness as it requires 'risk-taking' (Brooker and Lavers 2005a, 23, 194; Brooker 2010b, 149; 2009; Sorabji 2008; Shipman 2006) Tronson 2006). The cases also show that many judges are taking a harder line on assessing whether the parties' rejections of mediation are 'genuine' (Brooker 2013, 162, 161–166) and a key factor appears to be the savings to the court not the parties (Meggitt 2014; Suter 2015; Taylor 2014; Genn 2010). This forceful approach could lead to diminishing settlement rates as reluctant parties go through the 'motions' in mediation in order to satisfy court requirements (Brooker 2013, 161–166, Tully-Brunsdon 2009, 230–231).[66] It also places importance on the confidentiality of the process, as Meggitt (2014, 347–348) queries: how unreasonable must the parties' conduct be for a court review, which is governed by the rules of confidentiality and 'without prejudice' (Brooker 2013, 161; Bartlet 2015).

Confidentiality and 'without prejudice'

The importance of confidentiality is that the parties are able to discuss separately with the mediator in the 'caucus' what their settlement options are, examine the strengths or weaknesses of their cases, and use the mediator to communicate offers to the other party (Brooker 2013, 185–38; Bevan 1992, 19; Mackie et al. 2007; Kovach 2006, 429–439). The issue of confidentiality comes before the courts when the parties seek to enforce agreements reached in mediation or question the terms of the settlement or, exceptionally, when all parties waive their rights and invite the court to review the reasonableness of their offers in mediation (see for example; Ahmed 2010; Brooker 2010a,b; 2013; Koo 2011; Briggs 2009a,b; Kallipetis 2011; Cornes 2007).[67]

However, when challenges are made about confidentiality this attacks the basis of how mediation works and potentially weakens trust in the process (Cole et al. 2013, 16.6; Brooker 2013, 234–235; Kovach 2006, 438–9).

Without prejudice

Confidentiality 'overlaps' with the common law rule of 'without prejudice', which is given to statements made in negotiations (Brooker 2013, 254). The basis of 'without prejudice' is to encourage settlement by providing that admissions and concessions cannot be use in later litigation (see for example, Koo 2011; Altaras 2010). The courts recognise mediation is 'assisted negotiation', thereby extending 'without prejudice' to statements made while mediating[68] but there are a number of exceptions to this rule,[69] many of which have been applied to mediation (Allen 2008; Brooker 2009; 2010a, 2013, Cornes 2007; Koo 2011; Sorabji 2008; Zamboni 2003). Therefore, the court may permit statements which prove an agreement was reached in mediation[70] or that the parties 'acted reasonably' to 'mitigate their losses'[71] or to bring evidence which shows 'perjury, blackmail or other ambiguous impropriety'[72] such as duress[73] and, finally, statements that are 'without prejudice save as to costs'.[74] The other exceptions may in the future be extended to mediation which is to prove that an 'estoppel' exists, which a party was 'intended to act on and does in fact act'[75]; 'to explain delay or apparent acquiescence'[76]; to show 'whether the claimant acted reasonably as to his loss in his conduct'[77] and the newest exception which is 'to aid construction'[78] of an agreement.

Confidentiality

Farm Assist [2009] is the leading case on mediation confidentiality which exists either as an 'express' form through a contract to mediate or is 'implied' by 'analogy' with arbitration which affords confidentiality between arbitrators and the participants (Brooker 2013, Chapter 5; Koo 2011).[79] All the parties including the mediator must agree to waive confidentiality but this is not 'absolute' and evidence may be produced in court when 'it is in the interests of public policy', which in *Farm Assist* was allowing the mediator to be called as a witness in order to explore a claim of economic duress (Tumbridge 2010, 148; Brookes 2010a,b; Brooker 2013; Koo 2011).[80]

The courts have been reluctant to introduce special rules for mediator confidentiality based on the 'special' relationship that they have with the parties (Briggs 2009a,b; Brooker 2013, 204–208). The law on confidentiality in its current form has impacted in a number of ways on the practice of mediation, particularly as it has led contracts to be drafted to exclude not just the dispute that the court becomes involved in but matters that arise 'in connection with the mediation' and, furthermore, some contracts' clauses state that the mediator will not act as a witness (CEDR 2015; Tumbridge 2010, 148; Brookes 2009; Wood 2008a,b).

An investigation of lawyers' perspectives and practices relative to court-connected mediation

In England and Wales, mediation is now court-connected, which has had a substantial influence on the practice of the legal professions (solicitors and barrister), from the advice that they give to their clients in order that they are informed of their obligations under CPR, their role in assisting them in settling their case, or attending mediation and the professional training they must take. ADR knowledge is now required in professional training but many lawyers also undertake courses as mediation advocates to assist them in representing clients, which is a way of extending their legal practice (SRA 2014; Bar Council 2014; Brooker 2011). Agapiou's survey of construction lawyers found that 78 per cent had undertaken mediation training either through internal or external courses and 21 per cent had undertaken mediator training. Moreover, in the construction industry, lawyers are the leading profession in the provision of mediator services (Gould et al. 2010).

Research in the past 15 years suggests mediation is becoming more commonplace and many construction lawyers are repeat users (Brooker and Lavers 2005a,b; Gould et al. 2010, 10–11; Agapiou 2015; Sidoli de Ceno 2011). Agapiou's (2015, 240) survey shows that it is extremely unlikely that clients are a factor in the relatively low uptake of mediation, as 48 per cent of the lawyers reported that their clients 'never' or 'rarely' refused to use it. Nor is the lack of professional mediators affecting use, as this was found to have 'little' or no 'impact' on the decision (Agapiou 2015, 241). But what does influence lawyers selecting mediation is their knowledge and experience, which was found to correlate to their views on the efficacy of the process (Agapiou 2015, 239). One insight into the decision to mediate is that lawyers do not always 'initiate' mediation discussions 'regularly' and nearly half said that they 'never' or 'rarely' did unless 'compelled' by the court (Agapiou 2015, 238), which is likely to have a substantial impact on the timing of mediation, which is more likely to be after legal negotiations have failed and litigation is planned.

As with many other jurisdictions there are indications that some lawyers use the process in a more legalistic and adversarial way by presenting their clients' cases and using legal arguments to strengthen claims (Brooker and Lavers 2005a,b; Brooker 2007). Furthermore, evidence from various studies implies that more evaluative techniques are being used by mediators, which some lawyers and their clients prefer (Genn 1998; Brooker 2007).

The rise of evaluative mediation is often associated with court-connection and it is perhaps not surprising that many mediators in construction come from a legal background because disputes may be sizeable or very technical and parties are likely to have sought the advice of lawyers when negotiations break down who may then recommend lawyer-mediators (Gould et al. 2010, 10–11; Agapiou 2015).

The TCC survey reported that 75 per cent of appointed mediators were from the legal professions and a further 7 per cent were TCC judges (2010, 10–11). CEDR's Audit (2014, 4) indicates that over 50 per cent of respondents had legal qualifications but non-legal mediators were entering the profession, although there is a disparate level of experience between leading mediators, some of whom have lucrative careers and the majority of whom have not engaged in vast numbers of mediations, which suggests that mediator practice remains largely a secondary profession. The Audit (2014, 10–11) highlights that the most significant challenge to mediators is professionalizing practice with its expectation that it will increase work and bring better remuneration.

The Civil Mediation Council has taken the lead in the regulating practice by requiring members to satisfy benchmark requirements on training and continuing professional development (CEDR Audit 2014; Brooker 2011; 2013). There is, however, no obligation about what model or techniques mediators employ but if the mediator has not got a law qualification they must do a course in contract law (Brooker 2010, 2013, 252). A review of construction mediators' codes of practice suggests that the majority have adopted the EU Mediator Code but this does not constrain mediator styles, and an inspection of mediation contracts commonly used in England and Wales indicates that the process is in the hands of the mediators who should be guided by the parties or their lawyers but the rules are often prescriptive and legalistic in design (Brooker 2011, 38–39). Research in England and Wales has indicated that lawyers are instrumental in how mediation is practised, with the process becoming more evaluative with lawyer-mediators predominating who are often selected by repeat players. Future research should consider whether these developments are influencing practice and, if so, in which way.

Conclusions

Court-connection may have increased mediation but not as much as mediators or policy makers anticipated. CPR could be creating a more difficult environment for mediators to work in than previously encountered because rather than being commenced long before the parties become entrenched in legal arguments it frequently does not come into play until the parties decide to litigate after negotiations have failed, when the prospect of success may diminish the closer it comes to trial (Gould et al. 2010). The nearer mediation is to the trial, the more likely the procedure will reflect the arguments that have been rehearsed in legal negotiations, which increases the prospect that the parties' focus will be on legal positions and financial outcomes, which may make it difficult for mediators to nurture creative outcomes thus negating potential benefits. To what extent lawyers are responsible for the selection of the mediator has not been established but it is likely that the legal profession influences this decision when litigation is imminent. The majority of mediations are undertaken by lawyer-mediators (CEDR), which some evidence suggests is reflected in more evaluative practice, which can lead to a more prescriptive process where lawyer-advocates provide opening statements and drive settlement rather than the parties controlling events and having input in the final mediated outcome. Moreover, lawyers are more likely to recommend legally qualified mediators and more likely to feel comfortable in a process that they can influence, which will allow them to demonstrate their skills to their clients. Court rules have led to the connection of mediation to the litigation process but this has also resulted noticeably in a growth of mediation law, which is affecting decisions about electing to mediate, the timing of mediation, the expenditure and recovery of the costs of mediating, and how the process is conducted.

Notes

1 Civil Procedure Rule (CPR) 44.4(i) and (ii)
2 For an overview of the historical development of the 'modern mediation movement' in England and Wales see for example, Brooker (2013) Chapter 1; Brooker (2010), Genn 2010, Chapter 3; Mistelis (2006); Gould et al. 2010 Chapter II and III
3 Practice Statement, 10th December 1993 [1994]
4 CPR 1.4(2)(a)
5 under CPR s44(4)(a)
6 CPR 44.4(i) and (ii)
7 TCC Practice Direction 60.5.1
8 See *West Country Renovations Ltd v. McDowell and Anor* [2012] EWHC 307 (TCC)
9 See CPR Pre-Action Conduct 1.1
10 *Paul Thomas Construction Ltd v. Hyland* (2000) Adj. L. R 03/08
11 This protocol can be used in other courts when the dispute relates to construction or engineering.
12 TCC Protocol, 5.1 and 5.2
13 TCC Court Guide 2014, s2.4
14 TCC Court Guide 2014, 5.4
15 TCC Court Guide 2014, s7.2.3
16 TCC Court Guide 2014, s7.2.4
17 TCC Court Guide 2014, s7.1.1
18 TCC Court Guide 2014, s7.7.7
19 *Brookfield Construction (UK) Ltd v. Mott MacDonald Ltd.* [2010] EWHC (TCC) 659, 52–55
20 TCC Court Guide, 2014 s 7.6
21 TCC Court Guide, 2014 s 7.6.1
22 TCC Court Guide, 2014 Appendix G s4 and s8
23 *Halsey* [2004] EWCA Civ 576, 10
24 *Halsey* [2004], 30
25 Human Rights Act 1998 s6
26 See *Wright v. Michael Wright Ltd and Anor* [2013] EWCA Civ 234, 3
27 Ibid. at 3.
28 Ibid. at 3
29 Supreme Court Act 1981 s51; See *Chantry v. Convergence Group* [2007] EWHC 1774, 9
30 *National Westminster Bank v. Feeney* [2006] EWHC 90066, 20–21
31 *McGlinn v. Waltham Construction Ltd* [2005] EWHC 1410
32 *Lobster Group Ltd v. Heidelberg Graphic Equipment and Another* [2008] EWHC 157 TCC
33 *Roundstone Nurseries Ltd v. Stephenson Holding Ltd* [2009] EWHC 1431
34 TCC Court Guide (2014) s7.1.1
35 TCC Court Guide 2014, s7.2
36 *Alan Valentine v. (1) Kevin Allen (2) Simon John Nash (3) Alison Nash* [2003] ADR L.R. 07/04, 4
37 *Corenso (UK) Ltd v. The Burnden Group* Corenso [2003] ADAR. L.R. 07/01, 60
38 EU Mediators Code of Practice 2004, 3.2
39 *Wates Construction v. HGP Greentrue Allchurch Evans Ltd* [2005] EWHC 2174 (TCC), 29
40 See for example *PGF 11 SA v. OMFS Company Ltd* [2013] Civ 1288' *P4 v. Unite Integrated Solutions* EWHC 2924 (TCC) P4
41 *Halsey* [2004] 17
42 *Burchell v. Bullard* [2005] EWCA Civ 354; *Rolf v. De Guerin [2011]* EWCA Civ 78, 1
43 See *Brookfield Construction (UK) v. Mott MacDonald Ltd* [2010] EWHC (TCC) 65

44　*Halsey* [2004], 50; *Daniels v. Commissioner of Police for the Metropolis* [2005] EWCA Civ. 1312, 30

45　*Halsey* [2004], 18

46　Halsey [2004], 19

47　*Halsey* [2004], 20

48　*Northrop Grumman Mission Systems Europe Ltd v. BAE Systems (Al Diriyah C4I) Ltd* [2014]

49　*Garritt-Critchley and Others v. Ronnan and Solarpower* PV Ltd [2014] EWHC 1774 (Ch) 23–24

50　*Halsey* [2004] 22

51　*Sixth Duke of Westminster v. Raytheon* [2002] EWHC 1973; *Reed Executive Plc v. Reed Business Information Ltd.* [2004] EWCA Civ 887; *Nigel Witham Ltd. v. Smith and Anor* [2008] EWHC 12 (TCC); *S and Ors and Chapman and ANR (Chapman)* [2008] EWCA Civ. 800; *Corby Group Litigation v. Corby DC* [2009] EWHC 2109 (TCC)

52　*Reed* [2004], 35

53　*S and Ors and Chapman and ANR (Chapman)* [2008], 29–30

54　*Wates Construction v. HGP Greentrue Allchurch Evans Ltd* [2005] EWHC 2174 (TCC), 29

55　*Halsey* [2003], 23, 25–26

56　*Halsey* [2004], 19

57　*PGF 11 SA v. OMFS Company Ltd* [2013] Civ 1288, 29–30

58　*PGF II SA v. OMFS Co 1 Ltd* [2013] EWCA Civ 1288 29–30

59　*PGF II SA v. OMFS Co 1 Ltd* [2013] EWCA Civ 1288 30

60　*PGF II SA v. OMFS Co 1 Ltd* [2013] EWCA Civ 1288 30

61　*PGF II SA v. OMFS Co 1 Ltd* [2013], 34

62　*PGF II SA v. OMFS Co 1 Ltd* [2013], 35, 56

63　*Northrop Grumman Mission Systems Europe Ltd v. BAE Systems (Al Diriyah C4I) Ltd* [2014] EWHC 31 (TCC), 48

64　*Northrop* [2014], 59

65　*Northrop* [2014], 57

66　*PGF 11 SA v. OMFS Company Ltd* [2013] Civ 1288 45.2

67　*Carleton v. Strutt and Partner (A partnership)* [2008] EWHC 424 QB

68　*Aird v. Prime Meridian Ltd* [2006] EWCA Civ 1866, 5

69　*Unilever Plc v. The Proctor and Gamble Co* [1999] EWCA Civ 3027, 23

70　*Brown v. Rice and Anor* [2007] EWHC 625 Ch).

71　*Cumbria Waste Management Ltd (1) v. Baines Wilson (a firm)* [2008] EWHC 786 (QB)

72　*Hall v. Pertempt Group Ltd* [2005] EWHC 3110 (Ch), 13

73　See *Farm Assist Ltd. (FAL) v. Secretary of State for the Environment* (DEFRA) (No 2) [2009] EWHC 1102 (TCC)

74　*Reed* [2004], 21

75　Neuberger J in *Hodgkinson and Corby v. Wards Mobility Services* [1997] FSR 178, 191);

76　*Walker v. Wilsher* (1889) 23 QBD 335

77　*Muller v. Linsley and Mortimer* [1996] 1 PNLR 74

78　*Oceanbulk Shipping and Trading SA v. TMT Asia Ltd* [2011] 1 A.C. 662).

79　*Farm Assist Ltd* [2009] 44(1), 27, 44(1)

80　*Farm Assist Ltd* [2009] 28

References

Agapiou, A. (2015) The factors influencing mediation referral practices and barriers to its adoption: A survey of construction lawyers in England and Wales, *International Journal of Law in the Built Environment*, 7(3), 231.

Ahmed, M. (2010) Reinforcing the need to protect the without prejudice rule, *Civil Justice Quarterly* 29(3), 303.

Alexander, N (2006) *Global Trends in Mediation* 2nd edition, N. Alexander (ed.) Dordrecht: Kluwer Law International.

Allen, T. (2012) Don't ignore a request to mediate *Halsey* applied!:A note on *PGF II SA v. OMFS Company* [2012] EWHC 83 (TCC). CEDR Article, www.cedr.com/articles/?301. [Accessed on 6 July 2012].

Altaras, D. (2010) The without prejudice rule in England, *Arbitration* 76(3), 474.

Bar Council (2014–15) *Bar Standards Board Bar Professional Training Course* www.barstandardsboard.org.uk/media/1701167/bar_professional_training_course_and_covers_sept_15__revised_.pdf. [Accessed on 27 November 2015].

Bartlet, M. (2015) Mediation secrets 'in the shadow of the law', *Civil Justice Quarterly*, 112

Baruch Bush, R. and Folger, J. (2005) *The Promise of Mediation: The Transformative Approach to Conflict* (New and revised edition*).* San Francisco, CA: Jossey-Bass.

Bevan, A. (1992) *Alternative Dispute Resolution*. London: Sweet & Maxwell.

Blake, S., Brown, J., and Sime, S. (2013) *The Jackson ADR Handbook*. Oxford: Oxford University Press.

Boon, A., Earle, R. and Whyte, A. (2007) Regulating mediators? *Legal Ethics* 10(1), 26.

Brad Reich, J. (2002) Attorney v. client: creating a mechanism to address competing process, *Southern Illinois University Law Journal* 26, 183.

Briggs, Mr Justice (2009a) Mediation Privilege *New Law Journal*, 159(7363), 506.

Briggs, Mr Justice (2009b) Mediation Privilege *New Law Journal*, 159(7363), 550.

Brooker, P. (2007) An investigation of evaluative and facilitative approaches to construction mediation, *Structural Survey* 25(3/4), 220–238.

Brooker, P. (2009) Criteria for the appropriate use of mediation in construction disputes: judicial statements in the english technology and construction court, *International Journal of Law in the Built Environment*, 1(1), 82.

Brooker, P. (2010a) Mediation in the UK construction industry. In Brooker, P. and Wilkinson, S. (2010) (eds) *Mediation in the Construction Industry: An International Review*. London: Spon Press.

Brooker, P. (2010b) Judging unreasonable litigation behavior at the interface of mediation in the English jurisdiction, *Journal of Legal Affairs and Dispute Resolution in Engineering and Construction* 2(3), 148–161.

Brooker, P. (2011) Towards a code of professional conduct for construction mediators, *International Journal of Law in the Built Environment* 3(1), 24.

Brooker, P. and Lavers, A. (2002) Commercial lawyers' attitudes and experience with ADR4, *Web Journal of Current Legal Issues*.

Brooker, P. and Lavers, A. (2005a) Mediation outcomes: lawyers' experience with mediation, *Pepperdine Dispute Resolution Journal*, 2, 161–213.

Brooker, P. and Lavers, A. (2005b) Construction lawyers' experience with mediation post-CPR, *Construction Law Review* 1, 19–46.

Brunsdon-Tully, M. (2009) There is an A in ADR but does anyone know what it means anymore? *Civil Justice Quarterly* 28, 218.

Bucklow A. (2007) The 'everywhen' mediator: the virtues of inconsistency and paradox: the strength, skills, attributes and behaviours of excellent and effective mediators, *International Journal of Arbitration, Mediation and Dispute Management,* 73(1), 40.

Burnley, R. and Lascelles, G. (2004) Mediator confidentiality: conduct and communication, *Arbitration* 70(1), 28.

CEDR (2003), *Statistics,* www.cedr.com/press/?item=CEDR-Solve-mediation-statistics -2003. [Accessed on 27 April 2013].

CEDR (2014) *6th Mediation Audit,* www.cedr-asia-pacific.com/cedr/uploads/articles/pdf/ ARTICLE-20140603120956.pdf. [Accessed on 27 November 2015].

CEDR (2016), *Model Mediation Agreement,* www.cedr.com/about_us/modeldocs/?id=21. [Accessed on 27 November 2015].

Clark, B and Agapiou, A. (2011) Scottish construction lawyers and mediation: an investigation into attitudes and experiences, *International Journal of Law in the Built Environment* 3(2), 159–181.

Clark, B. (2012) *Lawyers and Mediation.* London: Springer.

Clark L.J. (Lord Clarke of Stone-cum-Ebony) The future of civil mediation, speech delivered to 2nd conference of the Civil Mediation Council, May 8, 2008. http://old. clerksroom.com/downloads/247-mr_mediation_conference_may08.pdf. [Accessed on 27 November 2015].

Cole, C, McEwen, C., Rogers, N. and Coben, P. (2013) *Mediation: Law, Policy and Practice.* London: Westlaw International Database.

Cornes, D. (2007) Commercial mediation: the impact of the courts, *Arbitration* 73(1), 12.

Dixon, G. and Carroll, E. (1990) ADR development in London, *International Construction Law Review* 7, 436.

European Union (2004) *EU Mediators Code.* http://ec.europa.eu/civiljustice/adr/adr_ec_ code_conduct_en.pdf. [Accessed on 26 April 2013].

Genn, H. (1998) *Central London County Court Pilot Mediation Scheme: Evaluation Report.* Lord Chancellor's Department, Research Series No 5/98. London: Lord Chancellor's Department.

Genn, H. (2002) *Court-based ADR Initiatives for Non-family Civil Disputes: The Commercial Court and the Court of Appeal.* Department of Constitutional Affairs (DCA) Research Series, 1/02. London: Department of Constitutional Affairs.

Genn, H. (2010) *Judging Civil Justice: The Hamlyn Lecture 2008.* Cambridge: Cambridge University Press.

Genn, H. (2012) What is civil justice for? Reform, ADR and access to justice. *Yale Journal of Law and the Humanities* 24, 397.

Gerber, P. and Mailman, B. (2005) Construction litigation: Can we do it better? *Monash University Law Review* 31(2), 237.

Gould N., King, C. and Britton, P. (2010) *Mediating Construction Disputes: An Evaluation of Existing Practice.* London: Kings College London, Centre of Construction Law and Dispute Resolution.

Gould, N., King, C. and Hudson-Tyreman, A. (2009) The use of mediation in construction disputes: Summary report of the final results, http://www.fenwickelliott.com/files/ Summary%20Report%20of%20the%20Final%20Results.pdf. [Accessed on 7 November 2010].

HMCS (2009) Her Majesty's Court Service, www.hmcourts-service.gov.uk/aboutus/ structure/index.htm. [Accessed on 13 February 2009].

Hensler, D. (2003) Our courts ourselves: How the alternative dispute resolution movement is reshaping our legal system, *Pennsylvannia State Law Review* 108, 165.

Hensler, N. (2001) Making deals in court-connected mediation: What's justice got to do with it? *Washington University Law Quarterly* 79, 787.

Imperati, S. (1997) Mediator practice models: the intersection of ethics and stylistic practices in mediation, *Willamette Law Review* 33, 703.

Jackson, L.J. (2009) *Civil Justice Costs Review The Final Report*. London: HMSO.

Judiciary website (2015) The Court of Appeal's mediation scheme, www.justice.gov.uk/courts/rcj-rolls-building/court-of-appeal/civil-division/mediation. [Accessed on 30 November 2015].

Kallipetis, M. (2011) Mediation privilege and confidentiality and the EU Directive, in Pointon, G. (ed.) *ADR in Business: Practice and Issues Across Countries and Cultures, Volume 2*. Dirdrecht: Kluwer Law International.

Koo, A. (2011) Confidentiality of mediation communications, *Civil Justice Quarterly* 30(2), 192.

Kovach, K. (2006) The evolution of mediation in the United States: Issues ripe for regulation may shape the future of practice. In Alexander, N. (ed.) *Global Trends in Mediation,* 2nd edition. Dordrecht: Kluwer Law International.

Kovach, K. and Love, P. (1996) Evaluative mediation in an oxymoron, *Alternatives to High Cost Litigation* 14, 31–32.

Kovach, K. and Love, P. (1998) Mapping mediation: The risks of Riskin's Grid, *Harvard Negotiation Law Review* 71(3), 109.

Mackie, K., Miles, D., Marsh, W. and Allen, T. (2007) *The ADR Practice Guide: Commercial Dispute Resolution*. London: Tottell.

Matz, D. (1994) Mediator pressure and party autonomy: Are they consistent with each other?, *Negotiation Journal* 10(4), 359–365.

Meggitt, G. (2014) PGF II SA v OMFS Co and compulsory mediation, *Civil Justice Quarterly*, 33(3), 335–348..

Ministry of Justice (2012) *Solving Disputes in County Courts: Creating a Simpler, Quicker and More Proportionate System: Consultation on Reforming Civil Justice in England and Wales* https://consult.justice.gov.uk/digitalcommunications/county_court_disputes/results/solving-disputes-in-cc-response.pdf. [Accessed on 27 November 2015].

Mistelis, L. (2006) ADR in England and Wales: a successful case of public private partnership, in Alexander N. (ed.) *Global Trends in Mediation* 2nd ed., Dordrecht: Kluwer.

Nolan-Haley, J. (1996) Court mediation and the search for justice through law, *Washington University Law Quarterly,* 74(1), 48–99.

O' Connor, P. (1992) ADR: Panacea or placebo? *Arbitration*, 58(2), 107–115 .

Phillips, L.J. (Lord Phillips of Worth Matravers) (2008) Alternative dispute resolution: An English view. www.judiciary.gov.uk/Resources/JCO/Documents/Speeches/lcj_adr_india_290308.pdf. [Accessed on 26 April 2013].

Riskin, L. (1996) Understanding mediators' orientations, strategies and techniques: A grid for the perplexed, *Harvard Negotiation Law Review*, 1, 7–51.

Rix, B, (2012) The interface of mediation and litigation, *Arbitration*, 80, 21–22.

Roberts, S. (2000) Settlement as civil justice, *Modern Law Review* 63(5), 739–747.

Shipman, S. (2006) Court approaches to ADR in the civil justice system, *Civil Justice Quarterly* 25, 181.

Sidoli del Ceno, J. (2011) An investigation into lawyer attitudes towards the use of mediation in commercial property disputes in England and Wales, *The International Journal of the Law of the Built Environment* 3(2), 182.

Sidoli del Ceno, J. (2012) Mediation and the judiciary: Negotiation is not enough *Nottingham Law Journal*, https://www.questia.com/library/journal/1G1-327955047/mediation-and-the-judiciary-negotiation-is-not-enough [Accessed 30 September 2016]

Sidoli del Ceno, J. (2015) Costs, mediation and the judiciary, *Arbitration* 81(1), 105–108.

Solicitors Regulation Authority (2012) *LPC Training Course Information*, www.src.org.uk/lpc. [Accessed on 26 April 2012].

Sorabji, J. (2008) Costs – further developments from Halsey: Nigel Witham Ltd v. Smith and Isaacs and S v. Chapman, *Civil Justice Quarterly* 27(4), 427.

SRA (Solicitor Regulation Authority) (2014) *Law Legal Practice Course*, www.sra.org.uk/students/resources/legal-practice-course-information-pack.page#heading_toc_j_14. [Assessed on 27 November 2015].

Suter, E. (2015) Unreasonable refusal to mediate and costs, *Arbitration*, 81(1), 2.

Taylor, P. (2014) Failing to respond to an invitation to mediate, *Arbitration*, 80(3), 470.

Technology and Construction Court (2013) *TCC Annual Report* www.judiciary.gov.uk/wp-content/uploads/2015/05/technology-construction-court-ar-2013–14.pdf. [Accessed on 26 November 2015].

Technology and Construction Court (2014) *TCC Court Guide* 3rd revision. www.justice.gov.uk/downloads/courts/tech-court/tec-con-court-guide.pdf. [Accessed on 26 November 2015].

Tronson, B. (2006) Mediation orders so the arguments against them make sense, *Civil Justice Quarterly* 25, 412–418.

Tumbridge, J. (2010) Mediators: confidentiality and compulsion to give evidence – issues in England, *International Company and Commercial Law Review* 21(4), 144.

Welsh, N. (2001a) The thinning vision of self-determination in court connected mediation: The inevitable price of institutionalisation? *Harvard Negotiation Law Review* 6, 1. http://papers.ssrn.com/sol3/papers.cfm?abstract_id=1724967. [Accessed 30 September 2016].

Welsh, N. (2001b) Making deals in court connected mediation: What's Justice got to do with it? *Washington University Law Quarterly* 79, 787.

White Book (2015) London: Sweet & Maxwell.

Wood, B. (2008a) Mediator privilege, http://billwoodmediation.com/articles/privelage.pdf. [Accessed on 16 April 2013].

Wood, B. (2008b) When girls go wild: The debate over mediator privilege, *The Mediator Magazine*, September.

Woolf L.J. (The Right Honourable The Lord Wolf, Master of the Rolls) (1995) *Access to Justice: The Interim Report*. London: HMSO.

Woolf, L.J. (The Right Honourable The Lord Wolf, Master of the Rolls) (1996) *Access to Justice: The Final Report*. London: HMSO

Zamboni, M. (2003) Confidentiality in mediation, *International Arbitration Law Review* 6(5), 175–190.

Zander, M. (1997) The Woolf Reforms: forward and backwards for the new Lord Chancellor, *Civil Justice Quarterly* 17, 208.

3 Court-connected mediation in South Africa

Olive Roma Caroline du Preez

Introduction

The introduction of mandatory mediation by the South African legislature raises the question of future trends relating to the practice of alternate dispute resolution (ADR) in the construction industry. The competitive and fast-track nature of the construction industry lends itself to an increased risk of dispute. Professionals invariably fulfil the role of project manager in the South African construction industry. With research suggesting that decisions are being made for the parties without negotiating their own settlement, the application of ADR is questionable in the South African construction industry (Povey, 2005). The inconsistent form of practice prevents the full realisation of the benefits of the advantages relating to the consensual methods of ADR. The question is, will current practice stand up to the challenges faced in regard to legislation and competitive practice and how will that influence future trends? The past and present situation relating to ADR practice in the construction industry was investigated in order to project possible future trends based on the evolution of the ADR process.

The history of ADR in South Africa

Arbitration has been used as a means of ADR since the days of Jan van Riebeek and was based on Roman-Dutch law. The present South African Arbitration Act 42 of 1965 and contract documentation are based on English Law. Mediation and conciliation in South Africa and the UK are in some respects applied in a different manner to internationally practised mediation (hereafter referred to as standard mediation practice). Mediation in the South African construction industry has created much confusion when comparing it to standard mediation practice. In standard mediation practice, the mediator is not required to recommend a solution to the dispute, but in the construction industry it is required of him. This has been addressed in the Joint Building Contracts Committee (JBCC) Contract documentation (1991) as the process of mediation has evolved in the construction industry.

Dispute resolution in the construction industry is different due to the use of unique adjudicative methods whereby judgements can be made with non-binding decisions which are characterised by consensual and control features. The roles of mediation and conciliation were originally reversed in the South African construction industry. The mediator was required to recommend a solution to the dispute while the conciliator fulfilled a facilitative role (Finsen 2005). It is important to note that the consensual nature of arbitration places it in the ADR context (Pretorius1993); Joint Building Contracts Committee Principal Building Agreement (JBCC PBA, (2014).

Agent resolution was included in the PBA of 1991 Dispute Clause 37, which was recommended by the JBCC (1991). Agent resolution as a method of dispute resolution was identified in the Association of South African Quantity Surveyors' (ASAQS) 1981 Practice Manual under the Agreement and Schedule of Conditions of Building Contracts (1981). Although agent resolution is not included in the JBCC contract documentation, it was included as a means of dispute resolution due to its popularity in practice (ASAQS, 1981; Verster and van Zyl, 2009; JBCC PBA, 1991).

Adjudication was introduced to the South African construction industry as an ADR method which requires expert determination and was included in the Joint Building Contracts Committee Principal Building Agreement (JBCC PBA) Series 2000 4th Edition (2004). This approach took effect after the change to adjudication, adopted from the Latham Report in the United Kingdom (Scott and Markram, 2004). Adjudication is applied as a dispute resolution method when the parties require a decision to be made for them; this decision is provisionally binding unless it is overturned in a subsequent arbitration (Finsen, 2005). South Africa adapted the method and created a less adversarial approach than that of the United Kingdom (Maritz and Hattingh, 2015).

Although conciliation is not included as a method of dispute resolution in the JBCC contract documentation, it may suffice as a method of informal dispute resolution as required in Clause 30.1 in the JBCC PBA (2014).

Mediation was introduced to the South African construction industry in 1976 as an alternative method of dispute resolution to the cost- and time-consuming method of arbitration (Quail, 1978). The Joint Study Committee issued a practice note with the intention of saving costs and time in the resolution of disputes. It was submitted that should parties be dissatisfied with the outcomes, they were still entitled to submit to arbitration (Quail, 1978). Despite the application of new methods to speed up the dispute resolution process, the unique and expeditious nature of the construction industry may lend itself to even more time-saving applications; hence the inconsistency indicated in Povey's (2005) research with the principles relating to an accepted mediation process and the evolution of a mediation process unique to the construction industry.

The similarities that exist between arbitration, the oldest method of ADR, and mediation may be appreciated because new methods were developed for the purpose of speeding up the arbitration procedure so as to provide a more informal and cost-effective way of resolving disputes most suited to the industry (Butler and Finsen, 1993). Besides arbitration, alternative methods of dispute resolution became more appealing in the management of projects when the rate of construction increased and the design and procurement of contracts became more complex (Finsen, 2005).

The practice of mediation has continuously been adjusted to suit the needs of the industry. The flexibility of mediation has placed it in a favourable position as a method of dispute resolution which may offer relief to court congestion in the justice system. Contrary to the flexibility of mediation practice, the adversarial nature of arbitration, and to a lesser degree, adjudication, has similarities relating to their litigation counterpart in dispute resolution.

Court-connected mediation has been proven to be successful in other countries, inter alia the United States of America, China and various European countries (Allen, 2013). The decision to implement court-connected mediation in South Africa was originally initiated by the previous Chief Justice in 2003. The court-based mediation rules were drafted and approved by cabinet in 2010 in an attempt to improve the justice system. One of the six purposes of court-based mediation is to 'facilitate an expeditious and cost-effective resolution of a dispute between litigants' (Department of Justice and Constitutional Development, 2014; Jordaan, [n.d.]). This initiative followed the successful implementation of the CCMA (Commission for Conciliation, Mediation and Arbitration) by the Department of Labour in 1995 (Benjamin, 2013).

Recent adjustments to the process of mediation have been made by the Department of Justice and Constitutional Development in the form of a Strategic Plan (2012–2017) in regard to court-connected-mediation rules which were drafted and submitted in December 2011 for promulgation and will be implemented gradually (Department of Justice and Constitutional Development, 2012). The relief of court congestion has been addressed by the Civil Justice Reform Project (2012) by simplifying lengthy and complex court processes and implementing ADR in the form of mandatory mediation in order to settle out of court. The intention was to implement these court-connected mediation rules gradually commencing in 2012. However, Joubert and Jacobs (2012) are of the opinion that for this process to be successful, the voluntary process of mediation would be considered an obstacle. This system may also be in conflict with the Dispute clause 30 of the JBCC 2014 which follows a voluntary process. Kuper (2016), chairman of the Arbitration Foundation of Southern Africa communicates mediation success rates to be relatively low at 20 per cent. However, this usually concerns large commercial disputes, which are heavily lawyered. Literature sources by Povey (2005), and Verster and van Zyl (2009) report mediation in the construction industry success rates at 80 per cent. Maritz (2009) suggests that despite mediation being reported as successful by Povey (2005) and Verster and van Zyl (2009), the trend is moving to adjudication. The implementation of the pilot project for court-connected mediation suggests that currently mediation has taken precedence over statutory intervention in the application of adjudication.

Court-connected mediation

The Department of Labour introduced the CCMA in 1998 to provide free justice for all. However, the CCMA did not deal with construction contractual disputes (Rwelamila, 2010). The only point of reference the construction industry had to ADR was the South African construction industry's Policy's White Paper. The White Paper on Creating an Environment for Reconstruction, Growth and Development in the Construction Industry (1998) encourages the use of dispute resolution in the South African construction industry to enhance the industry's performance.

Having followed the lead of the United Kingdom with regard to dispute resolution since 1976, South Africa continues to follow the trend. Specialists have been consulted on the implementation of the pilot project for court-connected mediation and approached to train mediators. Mediation rules have been developed and case management has been assigned to judges (Allen, 2013).

The commencement of the implementation of the court-connected mediation services was announced on 1 December 2014 by the Department of Justice in thirteen selected courts in the Northern areas of South Africa. Parties will be forced to participate or pay punitive costs (Henning, 2012). The intention is to move from an adversarial to a collaborative approach to dispute resolution, resulting in win-win situations and continuous business relations. Lawyers are apprehensive about court-connected mediation in fear of their billing hours being at stake. Nelson (2014) is of the opinion that the government's approach to court-connected mediation can only be successful if supported by lawyers.

The construction industry presents a challenge in the enforcement of court-connected mediation. There is a marked difference between the public and private sector, with the public sector supporting the industry with excessive contracts (Rwelamila, 2010). The Department of Public Works in South Africa has maintained the use of litigation as a means of dispute resolution (Finsen 2005). The private sector has created an informal type of dispute resolution in the form of conciliation, facilitated by professionals fulfilling the role of project manager. This is practised instead of direct negotiations as required in the JBCC PBA (2014) Dispute clause 40. Failing to lead the parties to a settlement, a dispute is declared. The construction industry operates on a contractual basis according to the various building contracts. Public works delete the Dispute clause and replace it with a litigation clause (Huggett, 2016).

Mediation has not displayed signs of progression since Povey's research in 2005, up to the point where court-connected mediation was introduced in 2012. Project-based mediation was not an alternative. However, Dispute Adjudication Boards (DAB) are considered to be effective for larger projects. The cost implications on smaller projects limit the use of this method (Swart, 2012).

Kuper (2016) is of the opinion that mediation will contribute to the decongestion of court roles, however: the process of mediation should be readily available to litigants, competently administered and expeditious. Kuper continues by suggesting that the process entails a properly resourced mediation facility, with effective case-handling staff and competent mediators. Kuper further suggests that South Africa's court system would find it difficult and expensive to deal with the workload and an inefficient mediation process will create delays and not improve the quality of justice. With regard to construction mediation, Kuper suggests that it is specialised and would require mediators with proper grounding in construction law and adequate experience. He doubts that such experience is available and that there are sufficient construction cases going through the courts to justify establishing a specialised mediation option. Kuper (2016) concludes that the construction industry has the best developed specialised ADR procedures and is not in need of external assistance.

Current ADR practice in the South African construction industry

The JBCC has been compiled to meet the needs of the industry in so far as it gives the parties a choice of a dispute resolution method they would prefer to use. With changing trends, the Dispute Clause of the JBCC PBA is often adjusted in the publication of a new edition (PBA, 1991; JBCC PBA, 2004; JBCC PBA, 2007; JBCC PBA, 2013; JBCC PBA, 2014).

The nature of construction mediation is conducive to conflict free negotiations which leads to advantages presented as the Four Cs. These advantages which constitute the features of the ADR context apply to the non-adjudicative methods of ADR and are as follows:

- consensus
- continuity
- control
- confidentiality
          (Loots 1991; Verster, 2006).

The ADR process is intended to give parties control and responsibility for the outcome (Bevan 1992). According to (Loots, 1991), these features produce effective outcomes.

Consensus

It is essential that consensus be reached between the parties, without which it would be an impossible task to facilitate or resolve a dispute (Bevan 1992). Consensus initially starts with consenting to the process/procedure. It suffices to say that the South African construction industry provides an opportunity for the parties to select their own method of dispute resolution in the JBCC PBA (2014); however, this is based on consensus on signing the contract. This only entitles the parties to submit to mediation without forfeiting their right to adjudication or arbitration. In view of the above, court-connected mediation provides for the basic fundamental of consensus allowing parties to agree on their choice of mediator in Rule 76.2a and Rule 77.4a (South Africa, Department of Justice, 2014).

Control

ADR practised in the construction industry allows for parties to be in control of the outcomes of a dispute resolution process. Apart from creating a win-win situation in regard to construction mediation, this suggests that through consensus both parties accept the outcome, thus creating outcome-based satisfaction (Moore, 1986).

Parties are self-empowered because they negotiate their own settlement and do not rely on a third party to make a decision for them, leaving them in control of the outcome (Brown and Marriott 1993). This form of conflict resolution may naturally leave the parties with a sense of control and empowerment which supports an environment of cooperation and involvement.

A skilled mediator is required to assist the parties to better understand the situation, to view the dispute on a broader context and to appreciate the other party's point of view. This makes it possible for the parties to make a decision based on the real issues to the dispute and allows them to be in control and satisfied with their decisions. Court-connected mediation complies with the standard basic fundamentals required in construction mediation with regard to control as indicated in Rule 80.1b, 1g (South Africa, Department of Justice, 2014).

Continuity

A continuous healthy business relationship is imperative in today's competitive construction industry. Loots (1991) suggests that irreparable harm to the ongoing business relationship should be avoided. Boulle and Rycroft (1997) and Moore (1986) suggest that mediation preserves and improves relationships by applying the 'gentle art' of reconciliation rather than the confrontationist process approach by the courts. Continuity between contractors and subcontractors is important because they depend on established relationships for performance of future contracts, hence the need for a cooperative attitude in the negotiation process (Finsen, 2005). Pretorius (1993) suggests that little harm can be done to a good existing relationship between the parties if the ADR process is managed effectively.

Court-connected mediation addresses continuity in the introductory 'purpose to promote restorative mediation' (South Africa, Department of Justice, 2014). The concept of continuity in business relationships is maintained.

Confidentiality

Confidentiality is important to the parties in respect of the integrity and ethics of their business. Facilitators need to regard confidentiality as a top priority in terms of withholding confidential information from the respective parties, in a relaxed and modest manner (Bevan, 1992).

Confidentiality is controlled by the disputing parties and no recordings and transcripts are made. Parties contractually commit themselves and any evidence which takes place behind closed doors is considered confidential and it cannot be used as evidence in a court of law (Boulle and Rycroft 1997; Trollip 1991). This is clarified in the first meeting. The JBCC, PBA Clause 30.1 (2014) states that if an agreement is reached it is put in writing and signed by the parties and considered final and binding however, it is still considered confidential.

Court-connected mediation generally ensures that the process remains confidential as supported by Rule 80.1.e (South Africa, Department of Justice, 2014).

The basic fundamentals of ADR

The language and techniques relating to the basic fundamentals of court-connected mediation are in place. In view of the arguments posed above with regard to the advantages relating to construction mediation, it would be beneficial to professionals/facilitators and contracting parties to be knowledgeable of these basic fundamentals of ADR, which lead to satisfactory end results. Research findings suggested that there was a certain degree of knowledge lacking which led to an analysis being conducted according to age groups. This determined that the advantages of conciliation are realised with experience. Lack of knowledge and understanding of the principles relating to the Four Cs may impact on the efficiency of construction mediation. Although the South African construction industry tends to adopt a hybrid form of ADR practice to suit the competitive nature of the industry, they do not entirely disregard standard international practice.

Arbitration

Arbitration is supported by most contract agreements, the Arbitration Act 42 of 1965, and the common law (Finsen, 1993). The arbitration clause was generally incorporated into a building contract and in the absence of this a dispute would be referred to a court of law (H.S McKenzie and S.D. McKenzie, 2009). According to Finsen (2005), arbitration has been the favoured method of dispute resolution for many years and is still considered an alternative method of dispute resolution to litigation because it offers more privacy and procedural flexibility. However, like litigation it is still based on court procedure and is of an adversarial nature (Finsen, 2005; Brown and Marriott, 1993). According to Finsen (2005), arbitration has become more formal with an improved decision making process. However, the cost and speed of arbitration have resulted in a move towards the more informal and speedy methods of dispute resolution. A positive aspect of arbitration is that it affords the parties the opportunity to select a decision maker with the appropriate expertise in construction. However, this may also apply to the other methods of ADR. Arbitration may therefore offer a competitive outcome as opposed to the satisfactory end result produced by mediation (Bevan 1992).

Although arbitration will, depending on the situation, be included as a method of the dispute resolution process, its win-lose nature (in so far as there are no negotiated outcomes and only an award (Finsen, 2005), may well impact on present and future relationships between the disputing parties, and as such the outcome could be measured in terms of present and long-term cost. Arbitration may be referred to as the 'backbone' of dispute resolution in the construction industry and forming part of the ADR context, because it may be considered a last attempt at resolving a dispute if other methods are unsuccessful. On failing to resolve a dispute in court-connected construction mediation, litigation may be considered to replace the function of arbitration in the construction industry contracts.

Agent resolution

In South Africa, the architect's discretion as principal agent was provisionally considered to be final and binding with the right to have it overturned by an arbitrator as a safeguard against biases. The employer and contractor may on consensual terms appoint the architect fulfilling the role of principal agent, as arbitrator (Butler and Finsen, 1993).

Architects were formerly given more authority than they presently have. However, the supervisory role is now assigned to the agent of the employer (McKenzie and McKenzie, 2009). An agreement as such may be to the advantage of the parties provided the principal agent remains impartial. The principal agent may be the most informed and qualified person on the issues of the project and be ideally suited to fulfil the role of mediator.

ADR research conducted by Verster and van Zyl (2009) indicates that agent resolution is a favoured method in the industry despite the fact that it does not offer the advantages offered by mediation.

Being a hybrid form of dispute resolution, agent resolution does not address all the advantages of construction mediation, however. It compares favourably with adjudication where a decision is made for the parties. As with adjudication and mediation, parties were permitted to submit to arbitration if they are not satisfied with the outcomes of agent solution (JBCC PBA 1991).The introduction of agent resolution illustrates the culture of the construction industry to adapt and develop methods of ADR to suit current practice. Pretorius (1993) refers to arbitration and adjudication as primary methods of ADR. All other ADR methods have stemmed from these.

Adjudication

A hybrid form of adjudication is practised in the South African construction industry. South Africa has a unique system of ADR where adjudication is adopted by the contracting parties (Finsen 2005). Adjudication in South Africa differs from adjudication in the United Kingdom, which is based on legislation and results in a final and binding decision (Maritz, 2009). In South Africa, adjudication is adopted by the contracting parties according to the agreement they have concluded (Finsen, 2005; Bevan, 1992). The use of adjudication was intended to speed up the resolution of disputes in order to avoid the loss of valuable contract time.

Clause 30 of the JBCC PBA (2014) entitles the parties to submit a dispute to adjudication, arbitration or to mediation at any time. Adjudication is supported by the Construction Industry Development Board (CIDB) and is now included in most construction agreements. Unlike the British method of adjudication where a binding decision is made, the decision in the South African construction industry is provisionally final and binding in so far as if the parties are not satisfied with the decision, it is subject to revision by an arbitrator (Maritz, 2009). Adjudication is not an obligation in the Dispute clause becaue parties are entitled to submit to arbitration or mediation (Finsen, 2005; Brown and Marriott, 1993; JBCC PBA, 2014).

The adjudicator acts as an expert; he/she receives the information on the dispute which is submitted by the parties and makes a decision (JBCC, 2007 4.1 Adjudication Rules, 6.3.1).Although adjudication is of an adversarial nature, sharing similarities with arbitration and litigation, it has become a well-used method of ADR in the construction industry. Adjudication or arbitration is an obligatory measure in the JBCC PBA (2007). However, Clause 40.5 states that it is not translated as a waiver of the parties to submit to mediation.

Consensus is practised in so far as the contracting parties agree on the method of adjudication and of the adjudicator. This may be agreed on at the time of drawing up the contract. Adjudication now takes precedence to arbitration in the dispute clauses of the following contracts: General Conditions of Contract for Works of Civil Engineering Construction (GCC) (2010), the New Engineering and Construction 3 (NEC3) (2005) and the International Federation for Consulting Engineers (FIDIC) (1999). Adjudication in the engineering discipline follows a different process in the form of Dispute Adjudication Boards (DAB) (Lalla and Ehrlich, 2012) which is supported by the FIDIC (1999), GCC (2010) and the NEC (2005) contracts, whereas in the building industry, adjudication is supported by the JBCC PBA (2014).

The South African Institute of Architects (SAIA) generally practise adjudication according to the JBCC PBA Dispute Clause 30 (The Cape Institute for Architecture, 2010). The DABs follow much the same process however; the establishment of the board differs where three adjudicators are appointed to resolve the dispute. The process relies on the expertise of engineers (Owen, 2003).

Maritz and Hattingh (2015) indicated that there is a growing preference for adjudication in the construction industry and courses have been developed in order to promote the practice of adjudication (Alusani, 2012). Maritz (2009) is of the opinion that adjudication will be underutilised without statutory intervention. The CIDB (2012) promotes adjudication as a rapid and cost-effective method. Adjudication has been practised according to the Adjudication Rules (2007) as an alternate method of dispute resolution to mediation and arbitration.

Adjudication enthusiasts were in favour of implementing the method of adjudication as a means of relieving court congestion. Maritz and Hattingh (2015) conclude that adjudication without statutory force is unlikely to succeed.

Conciliation

Conciliation in the construction industry equates to mediation in standard practice of ADR. Conciliation is the psychological component of mediation where the neutral third party will attempt to create an atmosphere of trust and cooperation. Conciliation is conducive to constructive negotiation. Conciliation also offers parties the opportunity to determine their own end results. Being a primary element of mediation, conciliation is applied with the intention of preparing the parties psychologically to enter into the extended evaluative process of mediation (Moore, 1986). As such, conciliation may serve as a preventative measure against disputes developing on site.

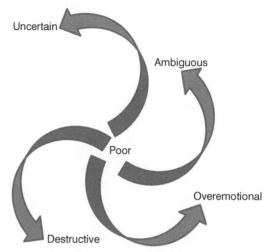

Figure 3.1 Unproductive communication (source: Adapted from Boulle and Rycroft, 1997)

Conflict invariably stems from poor communication and may inhibit the negotiation process. Ineffective communication may lead to a breakdown in the negotiation process, which may result in a deadlock. Poor communication skills create a negative cycle in which disputes are difficult to resolve (Richbell 2008; Boulle and Rycroft, 1997; Moore 1986). As illustrated in Figure 3.1, unproductive communication works against conciliation.

Ineffective communication presents a challenge in the facilitation of conciliation (Boulle and Rycroft 1997). Figure 3.2 illustrates the advantages offered by productive communication where positive outcomes are achieved (Boulle and Rycroft, 1997).

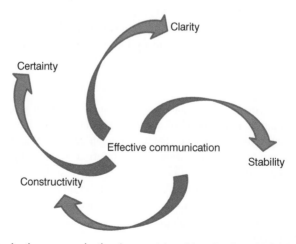

Figure 3.2 Productive communication (source: Adapted from Boulle and Rycroft, 1997)

As illustrated above, effective conciliation is supported by positive constructive negotiations for effective mediation.

Contemporary conciliation

Contemporary conciliation in the South African construction industry has evolved into a process which works for the industry. Upon failing to settle a difference between parties, a dispute is declared within ten days (JBCC PBA, 2014) of failing to resolve the subject of the difference. The route of dispute resolution is determined prior to concluding the contract. Failing to resolve the dispute, the parties will submit to mediation, adjudication, and arbitration. The above process largely applies to the private sector, while the public sector prefers to litigate despite the increased cost and time related challenges which may be incurred.

Mediation

In South Africa, standard mediation is primarily practised in labour law disputes and divorce matters (van Heerden and Potgieter, 2014). Mediation has been identified as a preferred method of ADR, however, the application of mediation in the South African construction industry has been questioned. Professionals are appointed as mediators due to their expert knowledge.

The mediation process is often expedited to suit the hurried nature of the construction industry. Mediators are reliant on inherent ADR communication skills and making decisions for the parties (Povey, 2005).

Mediation has slowly evolved over an extended period, culminating in a hybrid form of practice. It is clear that the intention was to identify a faster process in order to expedite the resolution of disputes. This process as illustrated in the JBCC Principal Building Agreements throughout the years, has been supported by the CIDB.

The informal procedures of mediation may provide a more favourable environment in which to effectively apply the Four Cs as suggested by Loots (1991). There is no set legislation for mediation and any natural person may facilitate mediation. Enforcement by the court of a settlement would be based on a contractual rather than a statutory provision (Finsen, 2005). The services of an expert mediator may be more appropriate for the construction industry. The emphasis which has been placed on the Four Cs highlights the advantages of mediation as opposed to the formal court system.

The only enforcement of the mediation process is in the contractual process wherein the JBCC PBA 2000 Edition 5.1 Clause 40.6.3 (2007) stipulates that on settlement a mediation agreement should be recorded and signed, upon which the mediation is considered final and binding. This is not stipulated in the mediation clause of the JBCC PBA (2014). However, the JBCC PBA 2000 Edition 6 (2013) has been slightly adjusted, requiring parties who settle before submitting to ADR to conclude a settlement in writing as required in Clause 30.1. This has been removed from the mediation clause of the JBCC PBA (2013).

When comparing research results by Povey (2005) and Verster and van Zyl (2009), there is a distinct preference for mediation and agent resolution, which may be regarded as an informal arbitration, referred to as being quasi-arbitral. This procedure, as practised in the construction industry, and as Dison (2006) suggests, should not be changed, however much it varies from standard practice.

Research conducted in the engineering field by Povey (2005) on mediation indicates that 24 per cent of the facilitators in the consulting engineering field are retired senior professionals, recalled to provide a mediation service. This may therefore have a correlation to the similarities of arbitration founded in the mediation process, hence the reference to mediation by Boulle and Rycroft (1997) as a quasi-arbitral function and by Dison (2006) as non-binding arbitration. In view of the mediation process being referred to as quasi-arbitral, in so far as arbitration principles are practised in mediation, this suggests that the practising arbitrators of the industry may also have moved on to practice mediation to keep current with a changing industry.

Despite the application of new methods to speed up the dispute resolution process, the unique and expeditious nature of the construction industry may lend itself to even more time-saving applications; hence the inconsistency indicated in Povey's (2005) research on the principles relating to an accepted mediation process and the evolution of a process unique to the construction industry.

Research indicates that the practice of mediation has been adjusted to suit the needs of the industry as reflected in the compilation of the various JBCC PBAs. According to the Principal Building Agreement 1991 edition and the General Conditions of Contract for Works of Civil Engineering Construction (GCC) published in 1990, the mediator was required to offer his opinion on the dispute. However, the opinion was binding if it was not rejected by the parties within a stipulated time (Finsen, 2005).

The JBCC PBA (2007) has no mention of the mediator expressing his/her own opinion and suggests that he/she not be too hasty to offer an opinion of a possible solution. Having a high regard for the mediator's authority and expert knowledge, parties may be inclined to request his/her opinion for a solution to settlement (Finsen, 2005). Published guidelines for mediation by the Association of Arbitrators and the South African Institute of Civil Engineers have no set rules of procedure, as one of the advantages of mediation is that the procedure should be flexible and left to the discretion of the parties (Finsen 2005). The GCC (2010) Dispute notice 10.3 replaces the functions of conciliation and mediation with the amicable settlement clause 10.4 and if not successful, it states that the dispute should be referred to adjudication within 14 days. This approach in the engineering field compares favourably with that of the JBCC PBA (2007). However, more emphasis is placed on mediation in the building industry.

Dison (2006) suggests that mediation in the South African construction industry is somewhat formal and has been described as non-binding arbitration. Boulle and Rycroft (1997) suggest that evaluative mediation is a quasi-arbitral function and the boundaries with arbitration are blurred. In view of the above, the method of conciliation may be best suited to the non-technical type of dispute where positive relationships are built, and mediation may be best suited to the technical type of dispute where expert advice and guidance is required.

When considering the differences between the conciliation and mediation methods of ADR, the extent of overlapping similarities are realized, as well as how difficult it is to clearly define the two methods. Brown and Marriott (1993) indicate that the distinction between conciliation and mediation is that conciliation tends towards a more facilitating approach whereas mediation tends to favour more of a proactive roll.

The evolution of mediation was somewhat dormant from 2006 to 2010 when adjustments to the procedure were considered. The vision of court-connected mediation of the Chief Justice of 2003 was rekindled in the form of the Strategic Plan of 2012–2017.

The standard practice of mediation compared to current practice

Table 3.1 compares the standard practice of mediation to the current practice as adopted in the South African construction industry. The table is divided into two columns. The left column is an illustration by Moore (1986) of the standard mediation process in a detailed description which works through set stages. It is important to note that the process may be somewhat tapered to suit the hybrid type of ADR practised in the South African construction industry, which is based on the conclusions of Povey's (2005) research. The reader may find that the stages relating to current practice have been adjusted or eliminated to more accurately reflect the situation in the industry. Due to the possibility of the lengthy stages being questioned in regard to the construction industry, the process was adjusted to accommodate the trend to expedite the mediation process.

Stages 2, 4 and 8 (see Table 3.1) regarding the strategy, design and hidden interests were eliminated due to the existing involvement of the professional fulfilling the role of principal agent or project manager. In standard practice, mediation may be conducted by a natural person. In the construction industry mediation is invariably conducted by an expert. The expedited hybrid form of mediation practice is illustrated.

Discussion

The practice of mediation is widely used in the construction industry with reference to the various contracts currently in use in the industry. The ADR process in the private sector is being applied effectively in the South African construction industry. The public sector may be a contributor of congestion in the justice system. Mediation as is currently practised in the construction industry challenges the process followed by the pilot project of court-based mediation. According to the JBCC PBA 2014, in the event of a difference developing between parties, they are required to apply direct negotiations or conciliation in an attempt to settle the dispute. In the event of failing to settle, the parties will be required to submit to mediation or any other form of ADR offered in the contract. Should they fail to

Table 3.1 The process of standard practice of mediation compared to current mediation practice in the construction industry

Standard practice of the mediation process	Current mediation practice in the construction industry
Stage 1: Initial contacts with the parties • Making initial contacts with the parties • Building credibility • Educating the parties about the process and selecting approaches	Stage 1: The dispute is reported according to JBCC PBA, 2014 Dispute Clause 30
Stage 3: Collecting and analysing background information • Collecting and analysing relevant data, dynamics, and substance of a conflict	Stage 3: The principal agent has all the information at hand
Stage 6: Beginning the mediation session • Opening negotiation between the parties • Establishing an open and positive tone • Assisting the parties in venting emotions • Assisting the parties in exploring commitments, salience and influence • The consensual approach to ADR in the South African construction industry is effective without statutory intervention in so far as it gives contracting parties the opportunity to use the method of dispute resolution best suited to their needs • The consensual approach to ADR in the South African construction industry is effective without statutory intervention in so far as it gives contracting parties the opportunity to use the method of dispute resolution best suited to their needs	Stage 6: Beginning the mediation session • Opening negotiation between the parties • Establishing the problem • Facilitator offers an opinion for a settlement
Stage 11: Final bargaining • Reaching agreement through either merging of positions, final steps to package settlements, development of a consensual formula, or establishment of a procedural means to reach a substantive agreement	
Stage 12: Achieving formal settlement • Identifying procedural steps to operationalize the agreement • Establishing an evaluation and monitoring procedure • Formalizing the settlement and creating an enforcement and commitment mechanism	Stage 12: Achieving formal settlement

Source: Adapted from Moore, 1986; Povey, 2005.

resolve the dispute, they would submit to arbitration. Without a contract, parties would be required to submit to court-connected mediation.

The language and techniques relating to court-connected mediation are in line with construction mediation as practised in the construction industry. However, secondary data suggests that there is a tendency for professionals to lack knowledge of the language and techniques in this field. Despite this, secondary data indicates that mediation is practised successfully in the South African construction industry (Povey, 2005). Existing pleadings and claim statements are required by the appointed court-based mediator. In the absence of these pleadings and claim statements, parties have a certain time period in which to submit (Gawanzura, 2014). This increases the risk of information shared with the mediator which can be used by the mediator in guiding the parties. Knowledge of intended claims may destroy the psychological component of trust and cooperation. Gawanzura (2014) suggests that this contradicts the way mediation is practised and may have a negative impact on the success rate.

Due to the technical nature of evaluative mediation in the construction industry, expert mediators are required to facilitate effectively. The project manager or principal agent is often required to fulfil the role of mediator. This may in all probability lead to cost savings with regard to the gathering of information which is already at hand. This is a cost saving which the pilot project is unable to accommodate. The question is, will the pilot project have available technical mediators?

Another cost-related challenge is the payment of mediator fees. In the past, payment of the mediation fees has been based on consensus between the parties. The pilot project states that the cost will be at the inviting parties' expense. A small contractor working on a large state project will be at risk with the cost implications of mediation fees which may be quite significant in large projects. In addition to this, corruption is a risk which may linger with the event of the appointment of mediators when the concept of consensus may be disregarded.

There is no doubt that court congestion will be relieved with the implementation of court-based mediation rules. However, the continuity which characterises the South African construction industry may be destroyed. The consensual approach to ADR in the South African construction industry is effective without statutory intervention in so far as it gives contracting parties the opportunity to use the method of dispute resolution best suited to their needs.

With regard to the public sector, construction professionals are apprehensive about the introduction of mandatory mediation. The choice and appointment of mediators may be politically influenced, creating a negative effect on the justice system. The construction industry is unique and qualified construction mediators will be required to make decisions. With regard to the private sector, construction professionals would prefer to continue with contemporary mediation. Construction professionals are faced with challenges with regard to the implementation of mandatory mediation. Although the intention is to relieve the current court congestion, the contemporary form of mediation will not coincide with the court-based system. Contrary to this, the same results relating to the relief of court congestion may be achieved.

One of the concerns in the 2014 report of the Minister of Justice, Mr. J.T. Radebe, is the training required for mediators. The report emphasises the lack of experience. The risk of providing effective mediation to suit the needs of a fast-track industry may be increased. The question is, will court congestion be relieved and will mediation be effective?

In addition to the United Kingdom, South Africa is following examples set by other African countries such as Nigeria and Rwanda. However, Gwanzuru (2014) is of the opinion that the implementation of court-based mediation needs to be applied correctly to be effective.

Allen (2013) is of the opinion that administrative procedures will have to be enhanced. This may result in further congestion. Further training may be required for effective implementation. Case management should be addressed by enhancing the authority of judges giving them case management powers and responsibilities. The question raised is, will the justice system manage these challenges once court-connected mediation has been implemented throughout South Africa?

The construction industry's contribution to the relief of court congestion is illustrated by the process followed by the private sector, which is contractually driven. However, the public sector may have to consider adjustments relating to the practice of mediation as their default method of dispute resolution. The question is, due to their previous record of success rates, will the private sector continue to practice construction mediation as before, or will they be forced into court-connected mediation?

Although the pilot project is considered a 'teething' process, South Africa may have a long way to go to achieve effective implementation.

Conclusion

The evolution of ADR in the South African construction industry has created a process which differs from standard mediation practice. Although the evolved process of mediation does not conform precisely to the pilot project of court-based mediation, it serves to relieve court congestion.

The practice of mediation has constantly been adjusted since the original introduction to mediation. As depicted in the comparison of standard with current mediation practice, the hasty nature of the construction industry lends itself to expediting the resolution of disputes rather than applying ADR according to the identified requirements. Measures may need to be taken to encourage professionals to apply the mediation rules in regard to the application of conciliation as an element of mediation.

ADR is continually being adjusted to keep trend with changes in the industry. This is illustrated in the regular adjustments made in each new edition of the JBCC Principal Building Agreements. In support of this, the Latham Report was adopted and adjusted at the time of its introduction in 2004. Professionals in the industry practice the Dispute Clause 30 in accordance to its requirements. As such it is suggested that the only way of ensuring that practice will be adjusted is by addressing the change in the JBCC PBA Dispute clause.

The research conducted on the contemporary practice of mediation suggests that that there are mediation facilitators who will need to expand their knowledge base of conciliation which forms the basis of mediation. Mediators have concentrated on the evaluative process of mediation with limited regard for the soft skills relating to conciliation. This may be supported by the experience mediators have in the construction industry in so far as the technical issues are addressed without feeling the need to consider the psychological component of conciliation. It suffices to say that mediation supported with effective conciliation may also contribute towards relieving court congestion. The hybrid form of ADR practice is successful. However, the use of the principles of standard mediation practice may add value to the process in order to eliminate the need to introduce new trends to settle out of court.

The construction industry will in all probability continue whichever practice suits their needs. The opinion of professionals in regard to the focus of mediation (with possible statutory intervention) may present challenges, as it is in conflict with the principles of the JBCC.

Looking back at the evolution of mediation in the South African construction industry suggests that this form of dispute resolution is well-suited and moulded to the industry. What is the point of changing a process which works? However, professionals have indicated that they are in favour of court-connected mediation, provided that the risk factors are managed; while others are adamant that it will not work. Professionals have been adapting to change since 1976. Will they find it difficult to adapt once more? A 'mixed feelings' scenario has been created: professionals may find it difficult to adapt to changing trends.

References

Allen, T. (2013). *Efficiency in Civil Justice and the Right Place for Mediation.* [Online] [n.p.] Available online at www.conflictdynamics.co.za/Sitefiles/206/Efficiency per cent20in per cent20civil per cent20justice per cent20and per cent20the per cent20right per cent20place per cent20for per cent20mediation per cent202014.pdf. [Accessed 15 January 2016].

Alusani Skills and Training Network (2012). *Construction Adjudication.* Available online at www.alusani.co.za/construction-adjudication. [Accessed on 22 November 2012].

ASAQS (Association of South African Quantity Surveyors) (1981). Agreement and Schedule of Conditions of Building Contract. In *Practice Manual.* Braamfontein: Association of South African Quantity Surveyors.

Benjamin, P. (2013). Assessing South Africa's Commission for Conciliation, Mediation and Arbitration (CCMA). International Labour Office Geneva Working Paper No. 47. [Online]. Available online at www.ilo.org/wcmsp5/groups/public/---ed_dialogue/---dialogue/documents/publication/wcms_210181.pdf. [Accessed 15 January 2016].

Bevan, A.H. (1992). *Alternative Dispute Resolution: A Lawyer's Guide to Mediation and Other Forms of Dispute Resolution.* London: Sweet & Maxwell.

Boulle, L, and Rycroft, A. (1997). *Mediation: Principles, Process, Practice.* Durban: Butterworths.

Brown, H.J. and Marriott, A.L. (1993). *ADR Principles and Practice.* London: Sweet & Maxwell.

Butler, D., and Finsen, E. (1993). *Arbitration in South Africa: Law and Practice.* Cape Town: Juta.

Cape Institute for Architecture (2010). *Note on Dispute Resolution in the Building Industry.* Cape Town: The South African Institute of Architects. Available online at www.cifa. org.za/UserFiles/File/PracticeNote052010.pdf. [Accessed on 26 September 2012].

Chartered Institute of Arbitrators. (2010). What is Alternative Dispute Resolution? Available online at http://asp-gb.secure-zone.net/v2/index.jsp?id=900/1127/2366&sta rtPage=25. [Accessed on: 3 September 2012].

CIDB (Construction Industry Development Board). (2012). *Provincial Stakeholders Liaison Meetings Report.* Available online at www.cidb.org.za/Documents/Corp/ SF_2012/SF_June2012_report.pdf. [Accessed on 21 November 2012].

Dison, L. (2005). An Investigation into the Mediation of Disputes in the South African Industry. *Journal of the South African Institution of Civil Engineering,* 48(1): 23.

Finsen, E. (1993). Arbitration and Mediation in the Construction Industry. In: Pretorius, P. (ed.).*Dispute Resolution.* Cape Town: Juta, pp. 176–206.

Finsen, E. (2005). *The Building Contract: A Commentary on the JBCC Agreements.* 2nd edn. Cape Town: Juta.

Gwanzura, M. (2014). Pilot of Court-Based Mediation Underway. Available online at www. iol.co.za/business/opinion/pilot-of-court-based-mediation-under-way-1.1792533. [Accessed on 2 January 2006].

Henning, M. (2012). Alternative Dispute Resolution in labour Management, Contract Negotiations and Generally in Unionized Environments has Deep Roots in SA. *Sunday Tribune.* Available online at http://wandahennig.com/2012/05/south-africa-opts-for-mediation-over-litigation. [Accessed 2 January 2016].

Huggett, D. (2016). Personal communication on court-connected mediation. Bloemfontein, 8 February 2016.

JBBC Principal Building Agreement 1991 (1991), Edition 5.0, Code 2101, JBCC Series (2000), June, Joint Building Contracts Committee, Johannesburg, South Africa.

JBCC (The Joint Building Contracts Committee). (1991). *Principal Building Agreement.* JBCC Series, Johannesburg: Joint Building Contracts Committee.

JBCC Series (2000) Edition 5.0, Code 2101, March, Joint Building Contracts Committee, Johannesburg, South Africa.

JBCC (2000) Principal Building Agreement & Nominated/Selected Subcontract Agreement, Ed. 4.1, cl. 1.1, 3.2, 6.0-7.0, JBCC Series (2000), March, Joint Building Contracts Committee, Johannesburg, South Africa.

JBBC Principal Building Agreement 2004 (2004), Joint Building Contracts Committee, Johannesburg, South Africa.

JBCC Adjudication Rules (2005) Adjudication Rules: for use with the JBCC Principal Building Agreement & Nominated / Selected Subcontract Agreement. March Edition, Joint Building Contracts Committee, Johannesburg, South Africa.

JBBC Principal Building Agreement 2007 (2007), Joint Building Contracts Committee, Johannesburg, South Africa.

JBBC Principal Building Agreement 2013 (2013), Edition 6.1, Code 2101, JBCC Series (2000), April, Joint Building Contracts Committee, Johannesburg, South Africa.

JBBC Principal Building Agreement 2014 (2014) Edition 6.1, Code 2101, JBCC Series (2000), April, Joint Building Contracts Committee, Johannesburg, South Africa.

Jordaan, B. [n.d.]. How to Prepare for a Mediation. Available online at http://capechamber. co.za/wp-content/uploads/2012/10/How-to-prepare-for-a-mediation.pdf. [Accessed 15 January 2016].

Joubert, J, and Jacobs, Y. (2012). Legal Brief Today. Available online at www.ens.co.za/news/news_article.asp?iID=577&iType=4. [Accessed on 21 November 2012].

Kuper, M.J. (2016). Chairman of The Arbitration Foundation of South Africa. Personal interview about mediation and arbitration statistics. Johannesburg, 7 March.

Lalla, N, and Ehrlich, D. (2012). A Guide to Using Dispute Adjudication Board. Johannesburg: Sonnenbergs. Available online at www.ens.co.za/news/news_article. asp?iID=577&iType=4. [Accessed on 27 August 2012].

Loots, P.C. (1991). Alternative methods of dispute resolution in the construction industry. *SA Builder* May, p 8–13.

Maritz, J.M. (2009). Adjudication of Disputes in the Construction Industry. *Essays Innovate, 3.* Available online at http://web.up.ac.za/sitefiles/file/44/2163/8121/Innovate per cent203/Inn per cent20bl78-79.pdf. [Accessed on 22 November 2012].

Maritz, J.M. and Hattingh, V. (2015). Adjudication in South African Construction Industry Practice: Towards Legislative Intervention. *Journal of the South African Institution of Civil Engineering*, 57(2): 45–49.

McKenzie, H.S., and McKenzie, S.D. (2009). *The Law of Building and Engineering Contracts and Arbitration.* Revised 6th edition by Peter A. Ramsden. Cape Town: Juta.

Moore, C.W. (1986). *The Mediation Process Practical: Strategies for Resolving Conflict.* San Francisco, CA: Jossey-Bass.

Nelson, A. (2014). Mediation Best Answer to Solve Conflict. *Cape Times*, 7 November Available online at www.iol.co.za/news/mediation-best-answer-to-solve-conflict-1777038. [Accessed 15 January 2016].

Owen, G. (2003.) *The Working of the Dispute Adjudication Board (DAB): Under New FIDIC 1999 (New Red Book).* Available online at http://fidic.org/sites/default/files/11%20DAB.pdf. [Accessed 30 September 2016].

Povey, A. (2005). An Investigation into the Mediation of Disputes in the South African Construction Industry. *Journal of the South African Institution of Civil Engineering,* 47(1): 2–7.

Pretorius, P. (ed.). (1993). *Dispute Resolution.* Cape Town: Juta.

Quail, G.P. (1978). *Die boukontrak.* (Vertaal en geredigeer deur R. Muller en L. Gaum). Pretoria: Boupublikasies.

Richbell, D. (2008). *Mediation of Construction Disputes.* Oxford: Blackwell.

Rwelamila, P.M.D. (2010). Construction Mediation in South Africa. In Brooker, P. and Wilkinson, S. (eds) *Mediation in the Construction Industry.* London: Spon Press.

Scott, R. and Markram, H. (2004). Adjudication: The Future in Alternative Dispute Resolution. Available online at www.deneysreitz.co.za/images/news/contructionmail2. pdf. [Accessed on 1 June 2011].

South Africa Department of Justice and Constitutional Development (2012). *Strategic Plan 2012–2017.* Pretoria: Government Printer.

South Africa Department of Justice and Constitutional Development (2014). *Rules On Court Based Mediation* (Version: March 2014). Pretoria: Government Printer. Available online at http://www.joasa.org.za/Mediation%20Rules%20Booklet_print-ready%20 FIN.pdf. [Accessed 30 September 2016].

Swart, B. (2012). Director of THM Engineers Pty Ltd. Bloemfontein. Personal communication on the use of DABs. 28 August, Bloemfontein.

Trollip, A.T. (1991). *Alternative Dispute Resolution.* Durban: Butterworths.

Van Heerden, C. and Potgieter, I. (2014). The Advent of Court-Based Mediation in South Africa. Available online at www.ggiforum.com/index.php/law/international-dispute-resolution/427-the-advent-of-court-based-mediation-in-south-africa. [Accessed on 2 January 2016].

Verster, J.J.P. (2006). Managing Cost, Contracts, Communication and Claims: A Quantity Surveying Perspective on Future Opportunities. In *Proceedings of 1st ICEC and IPMA Global Congress on Project Management,* 5th World Congress on Cost Engineering, Project Management and Quantity Surveying. Ljubljana, Slovenia, 23–26 April.

Verster, J.J.P. and van Zyl, C.H. (2009). Avoiding Differences and Disputes: A Construction Management Perspective. In *Proceedings of the 4th Built Environment Conference of Association of Schools of Construction (ASOCSA).* Livingstone, Zambia, 17–19 May.

4 Court-connected mediation in Hong Kong

Sai On Cheung

Introduction

The construction industry in Hong Kong has been experiencing strong growth in recent years due to major expansion of infrastructural developments. In 2014, the total gross value of main contract construction works increased by 12.5 per cent in nominal terms, amounting to HK$198 billion and by 5.9 per cent in real terms compared with 2013. Table 4.1 shows the contribution of construction industry to the gross domestic products (GDP) in Hong Kong.

Table 4.1 Contribution of the construction industry to GDP in Hong Kong

Year	GDP (HK$ million) in construction industry in Hong Kong	GDP (HK$ million) overall in Hong Kong	Percentage of contribution (%)
2004	42,142	1,316,949	3.2
2005	39,540	1,412,125	2.8
2006	40,591	1,503,352	2.7
2007	41,269	1,650,756	2.5
2008	51,225	1,707,488	3.0
2009	53,096	1,659,245	3.2
2010	58,619	1,776,332	3.3
2011	65,771	1,934,430	3.4
2012	73,334	2,037,059	3.6
2013	85,546	2,138,660	4.0

Source: Census and Statistics Department, Hong Kong, 2015.

Level of construction activity and number of disputes

The level of construction activity in Hong Kong is of unprecedented high as major infrastructural projects are under construction concurrently. Notably, ten 'mega projects' have been under construction to improve the transportation network within Hong Kong and in connection with that of mainland China. Although these projects have contributed to the economic growth of Hong Kong and will improve the well-being of Hong Kong people, it has been reported that most of these projects are experiencing cost overruns and schedule delay. In some of these projects, the press reported that completion costs would exceed the respective budget by over 60 per cent. The main reasons include increases in the project contingency costs, higher-than-expected tender returns and increases in labour and material costs.

Concurrent construction of these projects has worsened the already acute labour shortage. Issues including cost overrun and project delay could lead to construction disputes (Cheung, 2014). Therefore, the number of construction disputes is expected to rise in the next few years, in particular when these projects reach the near-completion stage. Having efficient and effective methods to settle these potential disputes is of prime importance.

Resolving construction disputes in Hong Kong

The Hong Kong International Arbitration Center (HKIAC) is an independent, nonprofit organization and is the principal provider of dispute resolution services in Hong Kong. Table 4.2 gives the record of dispute resolution cases managed by the HKIAC. In terms of dispute resolution method, arbitration remains the most used method. The number of mediations is relatively small. Adjudication is only occasionally used. As shown in Table 4.2, the total number of dispute resolution cases surged from 2006 to 2010. Since 2011, the annual total has been dropping. Moreover, as described in the last section, it is highly probable that the number of construction disputes will increase in the coming few years.

The practice of alternative dispute resolution (ADR) for construction disputes

The construction industry is a pioneer user of mediation. The use of ADR can be discussed at project level and when the dispute reaches the court system. Mediation is the most commonly used form of ADR for construction dispute resolution. Mediation is a voluntary, confidential and non-binding dispute resolution process in which the parties are assisted by a professional mediator to reach an amicable settlement. Unlike arbitration, a mediator does not decide the rightness or impose a settlement agreement. The mediator brings the parties together face to face in a private and confidential setting.

Each party has the opportunity of putting forward their point of view and listening to what the other party has to say (Cheung 2009). Although mediation is specified in the dispute resolution clause of the contract, it is not compulsory for

Table 4.2 Dispute resolution cases handled by HKIAC

Year	Total No. of dispute resolution matters	Arbitration cases	Domain name disputes	Mediation cases	Adjudication cases
2004	280	-	-	-	-
2005	281	-	-	-	-
2006	394	-	-	-	-
2007	448	-	-	-	-
2008	602	-	-	-	-
2009	746	429	140	159	18
2010	624	291	107	226	0
2011	502	275	127	100	0
2012	456	293	116	47	0
2013	463	260	170	33	0

Source: Annual reports of Hong Kong International Arbitration Center, 2015.

the parties. Under the HKIAC Mediation Rules, mediation can only be conducted if all the parties agree to mediate. Otherwise, failure by any party to reply within 14 days will be treated as a refusal to mediate. Mediation can be adopted by contract or agreement but if the detailed mediation procedures cannot be specified clearly in the contract, such clauses may not be enforceable. If, at the end of the mediation, both parties agree to the settlement arrangement and sign the settlement agreement, the parties are bound by the settlement contract.

Contractual use of mediation

In Hong Kong, most projects adopt standard forms of contract with necessary modifications to suit their use. It can be broadly said that mediation is typically used on a voluntary basis unless the dispute reaches the court.

In Hong Kong, most of the private construction projects adopt the building contract issued jointly by the Hong Kong Institute of Architects (HKIA), the Hong Kong Institute of Surveyors (HKIS) and the Institute of Construction Managers. The most up to date version is currently the 2005 Edition. Clause 41 is the dispute resolution clause and details a three-step dispute resolution procedure. Each party should specify one of its senior executives as its representative within 14 days of the acceptance of the tender. When a dispute arises, the architect, at the request of either party, shall first refer it to the designated representatives. The representatives of both parties shall meet and resolve the disputes within seven days of receiving the notice from the architect. If the dispute is not resolved by the designated representatives within 28 days, mediation shall then be conducted unless the dispute arises under Article 5. Such disputes involve documents forming the contract and other documents, which shall be referred to arbitration directly instead of mediation. The mediation shall be conducted in accordance with Hong Kong International Arbitration Center Mediation Rules unless otherwise agreed by the parties. After either party gives the counterpart a written request for mediation, the parties need to agree on the person to be the mediator within 21 days. Otherwise, the mediator shall be nominated co-jointly by the president or vice president of the HKIA and the HKIS. Where a dispute is not settled by mediation within 28 days of the commencement, either party can request the dispute be referred to arbitration by giving a notice to the other party. The arbitrator needs to be agreed between the parties. The arbitration generally shall not commence until after Substantial Completion or alleged Substantial Completion of the whole of the works or determination or alleged determination of the contractor's employment or abandonment of the works unless the written consent of the parties.

Most civil engineering works are funded by the Hong Kong government, hence the standard government form is the most commonly used conditions of contract for civil engineering works in Hong Kong. The use of mediation is very similar to the building contract as described in the preceding section. Recently, following the use of NEC3 by the governments in United Kingdom and Australia, the Hong Kong government has piloted its use through projects of the Drainage Services Department. The positive results from the pilot run has prompted the use of NEC (New Engineering Contract) Contracts for Government projects to be tendered in 2015/16 (NEC website, 2014).

The NEC3 suite of contracts adopts a two-step dispute resolution approach. The first step is to refer the dispute to adjudication. An adjudicator or adjudicator-nominating body can be named in the contract data. If no adjudicator is appointed, or he or she resigns or becomes unable to act, the parties may agree on an alternative or ask the adjudicator-nominating body to assign one. The referring party provides the other party with a notice of adjudication and refers the dispute to the adjudicator within seven days. The parties have 14 days to provide further information thereafter unless both parties have agreed an extension. The adjudicator then has 28 days to make a decision. The period can be extended by up to 14 days with the consent of the referring party. The adjudicator's decision is binding and final unless and until being revised by a tribunal and is enforceable as a contractual obligation between the parties and not as an arbitral award. Where either party is not satisfied with the decision or the adjudicator does not give the decision within time, either party can refer the dispute to the tribunal as long as he or she notifies the other party about his intention to refer to the tribunal within four weeks since receiving the adjudication decision. The final step is a tribunal that could be either arbitration or litigation. It is probably unexpected that NEC3 does not have any built-in participative type of ADR such as mediation given that NEC3 has been designed to promote cooperation and mutual benefits.

Mediation within the Hong Kong court system

In Hong Kong, the Construction and Arbitration List of the High Court is responsible for construction-related cases. With the aim of promoting wider use of mediation, the judiciary introduced a two-year pilot scheme for mediation of construction disputes in 2006. After the conclusion of the pilot scheme in April 2009, the Working Party reviewed the results of the pilot scheme and subsequently the judiciary decided to make the pilot scheme permanent through Practice Direction 6.1 (PD6.1). Under PD6.1, the classes of action within the list include but are not limited to all the cases concerning civil or mechanical engineering, building or other construction work, claims by or against engineers, architects, surveyors and other professional persons or bodies engaged in matters relating to the construction industry and applications relating to arbitration whether arising under the Arbitration Ordinance (Cap. 609), Rules of High Court ('RHC'), Order 73 or otherwise. Section F of PD 6.1 states that parties in construction are encouraged to attempt mediation and in order to promote the use of mediation, the court may impose cost sanctions where a party unreasonably refuses to attempt mediation. Cost sanctions may also be ordered if either party doesn't meet the committed minimum amount participation in mediation.

Section D of the Construction and Arbitration List stipulates that each party to a proceeding shall file with the court and serve on other parties the completed information sheet for which the solicitors of each party will be required to confirm whether they have informed their clients of the possibility of mediation to settle the disputes and whether the parties have tried mediation before. Where a party wishes to adopt mediation, he or she can do so by serving a Mediation Notice upon the other party and file the copy with the court. Upon receiving the Mediation Notice, the respondent should respond to the Applicant in writing with 14 days, stating whether he or she is willing to mediate, whether he or she agrees to the rules identified, whether he or she agrees with the timetable proposed and the minimum amount of participation the respondent proposes. Where the respondent refuses to mediate, he or she has to state the reasons as well. If differences exist in the timetables, rules or minimum amount of participation, the parties can apply to the judge to resolve the differences. Even if no party requests for mediation at the proceeding, the judge may ask the parties whether they have attempted mediation and reasons for not so doing.

There are generally two situations in which cost sanctions will be incurred for failure to attempt mediation.

1   Where a Mediation Notice is served, the other party refuses to mediate without reasonable explanations.
2   Where a Mediation Notice is served, the other party fails to engage in the mediation up to the minimum level of expected participation agreed by the parties or determined by the court.

PD 6.1 provides no definitive explanation on what constitutes a reasonable explanation for non-participation. However, it does state what the court will not consider in determining whether a party has acted unreasonably in refusing mediation. The court takes no account of what happened during mediation, the reasons for the mediation failure or any unreasonable conduct by any party

Table 4.3 Hong Kong cases involving refusal to mediate

Case	Brief fact	Salient points from the judgement
GOLDEN EAGLE INTERNATIONAL (GROUP) LIMITED and GR INVESTMENT HOLDINGS LIMITED HCA 2032/2007	The Defendant refused to mediate based on commercial reason & the belief that D has a strong case. The third explanation given is that D believed that the costs of mediation would be disproportionately high. The Court allowed COST SANCTION according to PD31.	1. Commercial reason is not a reasonable explanation. 2. The relief of a strong case identified by Dyson LJ is not suitable for Hong Kong cases. 3. This is a borderline case in which a party refused ADR because he thought that the would win should be given little or no weight by the court when considering whether the refusal was reasonable. Borderline cases are suitable for ADR. 4. There is no factual basis for the too high cost claim.
PACIFIC LONG DISTANCE TELEPHONE and NEW WORLD HCA 1688/2006 TELECOMMUNICATIONS LIMITED HCA 1688/2006	The Defendant refused to mediate on the ground that they both have previous unsuccessful attempts to settle this case. Therefore it was thought unlikely that mediation can help. The second explanation is that the Defendant submitted that the Plaintiff never made any proposal for resolution of the final account by mediation, which means the Plaintiff, has no genuine desire to settle it. No adverse cost is ordered to D.	1. The mere fact that negotiation failed to result in settlement does not mean that the parties would not benefit from mediation conducted by a skilled mediator. 2. The Court found that neither party made any moves towards settling those aspects through mediation. Notwithstanding the fact that parties put themselves at risk on costs by turning their back on mediation it is inappropriate to give any significant weight to the refusal to mediation.
SUPPLY CHAIN & LOGISTICS TECHNOLOGY LIMITED and NEC HONG KONG LIMITED HCA 1939/2006	The Plaintiff failed to file any reply to the Mediation Notice and requested a discontinuation of claims. The Court believed that it was an unreasonable failure but no order as to costs was ordered because claims were discontinued.	1. Usually before the Court suggests mediation, it would have already examined whether the case is appropriate for mediation. A party ignoring it shouldn't be surprised if the court seeks an explanation. 2. There is not much legal cost incurred due to the discontinuation of the claims.
GOOD TRY INVESTMENT LIMITED and EASILY DEVELOPMENT LIMITED DCCJ 3346/2011	The Plaintiff refused to mediate with the reasons as below: 1. The amount at stake is small. 2. The liability of D is clear. 3. The P has always been open for negotiation. 4. It would be disproportionate to claim and waste of money if the parties were to incur further costs. The Plaintiff accepted the Sanctioned Payment after refusal.	1. The outcome of mediation is always unknown. 2. Heavy litigation costs are expected in cases of such nature and they should be given serious consideration instead of the amount of the claim. 3. The judge doesn't agree with P's assertion that it has always been and open to negotiation because it was not the attitude as could be seen in the correspondence.

contributing to the failure in the course of mediation. Table 4.3 summarizes some recent cases involving refusal to mediate.

The Hong Kong Mediation Ordinance

The push for mediation in Hong Kong has been reinforced by the enactment of the The Hong Kong Mediation Ordinance (Cap 620) (MO hereafter). MO was enacted in June 2012 and came into effect on 1 January 2013. The objects of the Ordinance include: 1) to promote, encourage and facilitate the resolution of disputes by mediation; and 2) to protect the confidential nature of mediation communication. MO applies to any mediation conducted under an agreement to mediate. As the salient feature of mediation is allowing a private setting for settling disputes, confidentiality of the proceedings and the communication therein shall be protected.

In construction, mediation is commonly included as part of the dispute resolution clause. Such provision shall qualify as an agreement to mediate as far as MO is concerned, though most of these are mainly voluntary. Under Section 7 of MO, non-lawyers and foreign lawyers are allowed to participate in mediation as representatives and advisors. As mediation is meant to be a flexible process, MO does not regulate the arrangements of the proceedings. Instead, the mediation allows the parties to opt to be observed. In addition, rights and obligations that relate to mediated settlements are also not dealt with under MO. The parties have all the freedom to decide the legal form of the outcome of the mediation.

Thus, instead of procedural matters, MO ensures the confidentiality feature of mediation is kept. Section 8(1) of MO prohibits the disclosure of mediation communication. Under Section 2, mediation communication means (a) anything said or done, (b) any document prepared, or (c) any information provided for the purpose of or in the course of mediation, but does not include an agreement to mediate or a mediated settlement agreement. Moreover, disclosure may be permitted as provided in Section 8(2) and 8(3).

Under Section 8(2)(a), a person may disclose a mediation communication if the disclosure is made with the consent of(i) each of the parties to the mediation; (ii) the mediator for the mediation or, if there is more than one, each of them; and (iii) if the mediation communication is made by a person other than a party to the mediation or a mediator. If the content of the communication is already available to the public, Section 8(2)(b) allows disclosure. Disclosure is also possible if the content is otherwise subject to discovery in civil proceedings or to other similar procedures in which parties are required to disclose documents in their possession, custody or power. In this regard, it is generally taken as disclosure in the list of documents only, but not necessary for the purposes of production (Leung 2014). Under Section 8(3), disclosure is also possible with the leave of the court or tribunal.Another concern about the use of mediation is admissibility as evidence. Under Section 10, the court must take into account whether it is in the public interest or the interests of the administration of justice for admission of mediation communication to be disclosed or admitted in evidence in the Court of Appeal, the District Court, the Lands Tribunal and the Court of First Instance. MO provides a regulatory framework for mediation to be conducted in Hong Kong. Mediation and mediation communication are defined. The first ever MO in Hong Kong was a milestone in the development of mediation in Hong Kong.

Med-Arb provision under the Arbitration Ordinance (Cap 609)

The Arbitration Ordinance draws on the United Nation Model Law on International Commercial Arbitration and allows arbitrators to act as mediators prior to or following an arbitration with the consent of both parties by the 'Med-Arb provision'. Section 32(3) specifies that no objection can be made against the mediator acting as an arbitrator, or his conduct of arbitration, merely because the person had acted previously as a mediator in connection with some or all of the matters relating to the dispute submitted to arbitration, if the arbitration agreement has already specified that the mediator appointed will act as an arbitrator if no settlement is reached in the mediation proceeding. Section 33 further stipulates that the power of arbitrator to act as mediator as long as all parties consent in writing. An arbitrator may also act as a mediator even after the arbitral proceedings have commenced. Since all the information provided in the mediation process is confidential, the arbitrator acting as a mediator must treat the information obtained in mediation as confidential. Where the mediation proceeding terminates without reaching a settlement, the arbitrator must disclose as much of the information obtained in mediation as the arbitrator considers is material to the arbitral proceedings before resuming the arbitration.

The main advantage of Med-Arb provision is that it can reduce the costs and time in resolving the dispute by providing an arbitrator who has already had knowledge of the case to act as a mediator. On the contrary, a completely separate mediation by a mediator who has no knowledge of the case will cost much more and take more time to allow the mediator to gather information and explore the evidence. Med-Arb provision in the Arbitration Ordinance also gives the parties at the final stage of dispute resolution another chance of settling the dispute by reaching the settlement that both can live with at the stage of arbitration, which may largely reduce the costs and time spent compared to continuing with arbitration. And the parties can freely mediate because when the mediation fails, an arbitrator who is already familiar with the dispute can resume arbitration immediately.

However, one great concern about using Med-Arb provision that deters most people from trying it in Hong Kong is that, if the mediation is not successful, the confidential information that the arbitrator believes to be material will be disclosed despite the confidentiality of the information in mediation. This provision is unlikely to facilitate both parties exchanging all the information they have and having an open and honest communication to resolve the dispute, as they would be concerned that the information disclosed may be used against them at a later stage. As a result, the parties may hold back and refrain from disclosing their needs and compromises. The fact that the mediator or arbitrator has the right to decide which information is material is also widely criticized because this assessment is highly subjective. Another concern with Med-Arb provision is the risk of mediator bias towards the parties when he turns to act as arbitrator if the mediation fails. Even if some of the information or evidence submitted in mediation is not disclosed before resuming the arbitration, something said or done during the mediation may have made an impression on the arbitrator. Such an impression might influence the arbitrator's award. Thus, although Med-Arb provision may offer a more efficient and fast dispute resolution approach, it is very rarely used because of the concerns and disadvantages mentioned. The issues of confidentiality and potential bias need to be fully addressed and solved in order to promote such a provision and approach.

Court-connected mediation

It is believed that mediation can provide a cost-effective option to resolve disputes and avoid destroying relationships. Mediation facilitates engendering flexible solutions that both parties are happy to live with. Useful feedback on the use of court-connected and mandatory mediation can be found in Australia, United Kingdom and Singapore.

The Australian courts have successfully adopted court-connected mediation by both state and federal jurisdictions. Till now, the New South Wales, Victoria and Federal Courts have all enforced court-connected mediation through different Acts. Western Australia, South Australia and Tasmania also permit mandatory references despite any party's objection or refusal (Meggitt, 2008). Not only can the court order mediation without the parties' consent, but also the court can enforce any agreement or arrangement in the mediation session. In the court-connected mediations in New South Wales, a registrar or an officer of the court who is qualified as a mediator will conduct the mediation session. Parties cannot select which registrar will conduct the session. In 2010, the registrars of the court conducted 683 mediations with a success rate of 51 per cent[1]. Victoria has been and still is active in supporting mediation. In 1992, the Victoria Supreme Court started a 'Spring Offensive' to refer 280 cases out of 762 cases waiting for trails to mediation, which resulted in a 54 per cent success rate (Barlett, 1993). In 1994, the court commenced an 'Autumn Offensive', which referred 150 cases to mediation and almost 80 per cent of them were resolved successfully (Golvan 2000). At the same time, the Department of Justice assessed the use and effectiveness of mediation in the Supreme and County Court of Victoria and launched the report in 2009 (Hilmer, 2012). The report found that mediation meets all of the core objectives for ADR. The report of the Dispute Settlement Centre of Victoria (DSCV) – a free dispute resolution service funded by the Victorian Government, shows that 85 per cent of disputes are resolved at mediation and 86 per cent of DSCV clients are satisfied with mediation sessions in particular preserving existing relationship and enabling future business. Confidentiality and flexibility are two attributes of value to the disputants. In addition, the Federal Court's ADR programme has been routinely referring cases to mediation.

From 2013 to 2014, there were 567 cases referred to mediation out of 5009 total filings – this represents an 11 per cent referral rate, while the rate of referrals of applicable filings[2] is 23 per cent. Among all the cases referred to mediation, 67 per cent of them were resolved in full or in part. There is almost no difference in the success rates or user satisfaction between mandatory or voluntary mediation. A mediation ordered by a judge removes the concern that requesting mediation is a sign of weakness. [will make the concern that requesting for mediation will show weakness unnecessary.]

Though the UK does not adopt court-connected mediation widely at the moment, it has several pilot schemes to test the outcomes and user experiences of referring disputes to mediation. Among them, a quasi-compulsory mediation program 'Automatic Referral to Mediation' (ARM) was conducted by the Central London County Court between 2004 and 2005 after seeing the great success of the Ontario experiment in mandatory mediation. This pilot aimed to randomly refer 100 dispute cases per month to mediation, but the disputants were given the chance to object and opt out. A district judge reviewed cases in which objections were raised and could persuade (rather than require) the parties to proceed to mediation. This is because there was a ruling in 2004 that compulsory mediation ordered by the court might infringe access to a fair trial. As a result, there were only 172 cases actually mediated in this pilot. 81 per cent of the cases received objections and a small number of them still went to mediation after the judge's persuasion. This high opt-out rate makes this pilot very different from the Ontario experience in which the exemption rate was between 1 and 2 per cent, and could hardly be described as mandatory mediation pilot programme. It is of interest to note that where neither party raised objections initially, the settlement rate was 55 per cent, while this rate dropped to 48 per cent for those cases where objections were initially raised. This could hint that the high success rate of voluntary mediation might not be applicable to even just quasi-compulsory mediation, not to mention wholly mandatory mediation processes (Genn et al. 2007).

Both mandatory court-connected mediation and non-mandatory mediation are commonly adopted in Singapore. The State Court has been implementing mandatory mediation after a pilot programme started in 1994. From 1996, Court Dispute Resolution – one of the pre-trial stages that provide ADR services – has been integrated into the trial process. Under O34 of the Rules of Court, the court has the power to make necessary orders for the 'just, expeditious and economical disposal of the cause or matter' at pre-trial conferences, a compulsory step in the pre-trial stages. In 2012, 'presumption of ADR' was introduced. All cases would be referred to ADR automatically during summonses for directions or pre-trial conferences, unless parties withdrew from ADR. With the implementation of the Presumption of ADR, the Primary Dispute Resolution Centre has been handling more than 6,000 cases every year, with a successful rate of 85 per cent (Quek, 2015).

As the courts advocate the use of ADR as the first stop in resolving disputes, cases under particular legislation or subsidiary legislation will be submitted to mediation. For instance, under the Women's Charter (Cap 53), the court might refer the parties to mediation after considering the possibility of a harmonious resolution of the matter and with the consent of the parties.

The Supreme Court might have richer experience in directing cases to mediation through its case-management process. This is also supported by Justice Judith Prakash who once mentioned in her speech that the mediation as a diversionary approach/tactic is the resolution of disputes between parties and is actively carried out at pre-trial hearings (Prakash, 2009). In 1996, a pilot project, the Commercial Mediation Service, began during pre-trial conferences, parties are strongly encouraged to refer the case to mediation by the registrar, who will explain the process, advantages and costs of mediation to the parties. Within a year, 84 cases were referred to mediation with the settlement rate reaching 75 per cent (Quek, 2015). Recently, the concept of ADR offer was introduced. Although the act of making an ADR offer is voluntary, refusal of the offer may lead to the imposition of cost sanctions pursuant to O59, Rule 5(c) of the Rules of Court, which provide that:

> The Court in exercising its discretion as to costs shall, to such extent, if any, as may be appropriate in the circumstances, take into account ... the parties' conduct in relation to any attempt at resolving the cause or matter by mediation ...

Starting from 31 December, 2013, 2,291 cases have been referred to the Singapore Mediation Centre, which provides private commercial mediation and other related services to facilitate the resolution of disputes before trial. Among the cases directed, 73 per cent were settled successfully (Onn and Koh, 2015). As such, both mandatory and voluntary use of mediation are achieving similar settlement rates.

Mandatory use of mediation in Hong Kong: views from stakeholders

In Hong Kong, the mandatory use of mediation was deliberated by the Working Party on Mediation, which was established to consider how mediation could facilitate civil disputes in various levels of courts and measures to promote mediation. The Working Party published an Interim Report in 2001 incorporating six proposals for mediation in dispute resolutions. They are: (a) mandatory mediation in defined classes of case unless exempted by court order; (b) mandatory mediation by the court's order and proceeding stayed; (c) mandatory mediation by election of one party; (d) mandatory mediation as a condition of legal aid for appropriate types of cases; (e) unreasonable refusal to mediate reflected in adverse cost orders; (f) purely voluntary mediation with adequate information provided. These proposals are to help to discuss to what extent mediation shall be brought into the civil justice system. In the consultation stage, only the proposal (f) received general support.

In the Final Report (Department of Justice 2004), the Working Party rejected proposal (a) because mediation by statutory rule is an inflexible rule that will automatically direct cases that are unsuitable for mediation to mandatory mediation process. And this would necessitate a procedure of the application for exemptions, causing additional inconveniences. Proposal (c) was also rejected on the ground that allowing one party to impose its will on another party without any judicial control (decision from the court) would worsen the relationship and make the mediation less likely to succeed. Besides, it is likely to become a tactic for one party to delay proceedings. Proposal (b) attracted a number of objections, including voluntariness, duty to entertain litigation, lack of infrastructure and additional cost points, all of which were not accepted by the Working Party. Faced with the wide public concern that parties being forced to mediate will not be truly cooperative, the Working Party responded that court-connected mediation only requires participation and settlement may or may not be reached as long as a proper mediation has been conducted by a skilled mediator. However, this proposal was still rejected because the Working Party thought that it is not ready for Hong Kong to adopt court-connected mediation by court orders at this stage.

Proposal (d) was recommended by the Working Party because in the Working Party's view, this proposal is not discriminatory against poorer litigants but is just to provide a likely means of achieving a satisfactory resolution and of saving public resources. Where the mediation fails, or where mediation is inappropriate, the funding for court proceedings will still be provided for litigants. The director should have the power to require a claim to be pursued in the most cost-effective manner available and to revoke a legal aid certificate where the aided litigant 'has required the proceedings to be conducted unreasonably so as to incur an unjustifiable expense to the director or has required unreasonably that the proceedings be continued' according to the Legal Aid Ordinance.

Proposal (d) is also preferred by the Working Party. It is possible that a request for mediation can be made by a request from one party or the court's recommendation. At the same time, in order to relieve the anxieties about the confidentiality of mediation, the court should not inquire into how or why an attempt at mediation failed. Only an unreasonable refusal to proceed to the required degree of participation shall be taken into consideration. In deciding on the reasonableness of the refusal to mediate, the six elements proposed in the case *Halsey v. Milton Keynes General NHS Trust* shall be taken into consideration and they are (1) the nature of the dispute; (2) the merits of the case; (3) the extents to which other settlement methods have been attempted; (4) whether the costs of ADR would be disproportionately high; (5) whether any delay in setting up and attending ADR would have been prejudicial; and (6) whether ADR had a reasonable prospect of success. It is observed that the decisions in case laws are sometimes not consistent. In most cases, the explanations that one party has a watertight case won't justify the refusal but in the case of *McCook v. Lobo* [2002] EWCA Civ. 1760, the judge found that the issues were narrow and that the defendants had been found to have a strong case can justify the refusal. But the Working Party believed that these cases would still give great help to the Hong Kong courts notwithstanding the inconsistencies.Besides the recommendation of both proposal (d) and proposal (e), since the Working Party only rejects proposal (c) because it is likely to raise doubts at present, and it still expects that rules can be adopted to empower the courts to make more demanding mediation orders after

Table 4.4 The pros and cons of mandatory mediation

Proposals	Responses received during consultation
Rules making mediation mandatory in defined classes of cases, unless exempted by court order, should be adopted. A rule should be adopted conferring a discretionary power on the judge to require parties to resort to a stated mode or modes of ADR, staying the proceedings in the meantime. A statutory scheme should be promoted to enable one party to litigation to compel all the other parties to resort to mediation or some other form of ADR, staying the proceedings in the meantime. Legislation should be introduced giving the Director of Legal Aid power to make resort to ADR a condition of granting legal aid in appropriate types of cases. Rules should be adopted making it clear that where ADR is voluntary, an unreasonable refusal of ADR or uncooperativeness during ADR process places the party guilty of the unreasonable conduct at risk of a costs sanction. A scheme should be introduced for the court to provide litigants with information about and facilities for mediation on a purely voluntary basis, enlisting support of professional associations and other institutions.	**Five broad concerns:** 1. The imposition of any requirement to engage in mediation as a condition of being allowed to proceed with litigation is inconsistent with the right of access to the courts guaranteed by the Basic Law art. 35 and so is unconstitutional. 2. The court should perform its duty to hear cases in the usual way and should not direct or encourage parties to go elsewhere to resolve their dispute. 3. Hong Kong does not have the necessary infrastructure to adopt a court-connected ADR or mediation scheme. 4. Mediation must, by its nature, be voluntary and mandatory schemes are inherently likely to fail. 5. Such schemes are likely often to be counter-productive in that mediation, which fails, adds to the costs and delays. **Two specific objections:** 1. That the 4th proposal is objectionable since it is discriminatory against poorer litigants who have to rely on legal aids; and 2. That the 5th proposal suffers from the defect that no workable method of deciding whether a party has acted unreasonably or uncooperatively exists. And moreover, that any attempt to examine why a mediation or other ADR process failed, would impair the confidentiality and without prejudice nature of such processes essential to their success.

Source: Department of Justice of HKSAR, 2004.

more experience and expertise is gained, such mandatory mediation by the court's discretionary power is likely to take effect in the future. Table 4.4 summarizes the proposals and the key concerns.

At the time of writing, Hong Kong does not have court-connected mediation. Both Practice 6.1 (Construction Disputes) and Practice 31 (General Civil Disputes) stipulate that an unreasonable refusal to mediation when there is a Mediation Notice served may lead to adverse cost orders. However, a cost order will not be made against a party who has engaged in a minimum level of participation in mediation. Where a party has a reasonable explanation for not engaging in mediation, an adverse cost order will not be sanctioned. In 2010, the Working Party published a report on mediation stating that the development of mediation is still at an early stage and that it will be desirable to wait for the impact of civil justice reforms (CJR) on the use of mediation in Hong Kong to be analyzed. The Working Party also recommends revisiting the question of compulsory mediation in the future rather than introducing it without a consensus view from the stakeholders. Starting in 2010, the Hong Kong judiciary required parties of construction cases who had conducted mediation to report the particulars of the mediation. This feedback may help and inform the judiciary on how to proceed with the use of mediation for construction cases within the court system.

Is mediation a suitable ADR for construction disputes?

On the subject of resolving construction disputes through the use of ADR, is mediation the most suitable form? Litigation is considered the natural forum for settling disputes. For example, public policy issues may require a legal mandate and thus should be litigated rather than settled. Moreover, it is usually used as the last resort. Understandably, the majority of lawsuits are suitable for settlement before costly pre-trial activities and trial dates are set. The use of mediation can be considered at a much earlier stage. Whether a case is suitable for mediation can be considered by the nature of the dispute, the merits of the case, whether the costs would be disproportionately high and whether there is a reasonable prospect of success. Most disputes arising out of construction projects are boiled down to monetary issues; although legal points may be involved, argument over crucial points of law is not the main objective.

In the UK, the construction industry is one of the largest users of mediation (Nesic, 2001). In *Burchell v. Bullard [2005] EWCA Civ. 358*, the court stressed that a building dispute is 'par excellence' the kind of dispute that lends itself to ADR and the merits of the case favored mediation (Genn et al., 2007). Judge Robert Carmine Zampano (1991) also stated a similar view in the chapter Court-connected ADR: A View from the Bench, included in the book *Alternative Dispute Resolution in the Construction Industry* (Page 228):

> In 26 years as a federal judge, I can think of no large construction dispute that was not suitable for ADR. Fortunately in Connecticut, almost all complex construction litigation has been resolved before a trial and, most times, before commencement of discovery.

Feinberg (1991) also stated that mediation could offer a non-adversarial approach to foster good will and develop reputable relationships.

Construction disputes usually involve lengthy and extensive discovery, interlocking claims and counterclaims, complicated technical issues and meticulous interaction among professionals and experts (Zampano, 1991). Long and costly litigation for construction disputes will result in unpleasant results for the disputants. When disputants try to identify settlement possibilities in negotiations, a neutral mediator having no interest in the dispute would be more likely to be able to facilitate a settlement arrangement. Self-generated Settlement is preferred over court when ongoing business relationships are to be maintained and future dealings are anticipated. For various current or future construction projects, and to minimize the damage that could be done to the relationship or the ongoing projects at hand, a settlement agreed and satisfied by both parties will be much better than a decision made by a third party. Construction disputes may arise at any time during the construction period, with arbitration generally only being conducted after the practical completion; mediation would be an amicable alternative during the course of construction. Disputants would need 'emotional venting' and 'anger release' before getting a rational settlement. In this regard, mediation sessions would be a more amenable forum than courtrooms. It is also more likely that the parties will reveal their true 'bottom line' to a mediator rather than directly to the counterpart in a negotiation (Gillie, 1991).

Another challenge is the question of certainty of mediation clauses in enforcing contractual clauses. Many construction contracts include negotiation and mediation clauses in the hope of resolving most disputes before formal proceedings such as arbitration and litigation. Therefore, these clauses can lay down several steps that parties agree to undertake before either of them can go to litigation. However, inclusion of such clauses can lead to satellite litigation on the validity and enforceability of these clauses. One question is that, since such clauses are part of the contract, where the contract is invalid or non-existent, will the mediation clause can be severed like an arbitration clause (Hilmer, 2012)?There is no legislation to support mediation clauses like the Arbitration Ordinance that underpins arbitration. The enforceability of mediation clauses is decided based on the general contractual principles (Hilmer, 2012). In *Walford v. Miles* [1992] 2 AC 128, the court decided that 'the agreement to negotiate is unenforceable because it lacks the necessary certainty' (Lees, 2015). An agreement to negotiate is not enforceable if the obligations cannot be determined with sufficient certainty and if compliance or otherwise can be objectively assessed. Nevertheless, in *Channel Tunnel Group Ltd v. Balfour Beatty Construction Ltd* [1993] AC 334, this view was weakened since the court decided that it has a discretionary power to stay proceedings if the dispute resolution clauses are considered equivalent to an effective agreement to arbitrate. In the Hong Kong case of *Kenon Engineering Ltd v. Nippon Kokan Koji Kaisha* [2003] HKCFI 568, though the parties have agreed on the mediator in the contract, the judge decided that the clause is still not clear because the extent of matters to be mediated as well as the time limit for mediation haven't been agreed on in the contract. In this case, the contract states that disputes in relation to Sub-Contract or breach thereof will be mediated. In *Hyundai Engineering & Construction Co Ltd v. Vigour Ltd* [2004] HKCFI 205, the court believed that agreement to mediate provisions were unenforceable because they lacked detailed procedures and certainty. Another legal fallacy of such clauses is the inclusion of 'agreement in good faith' (Hilmer, 2012). The traditional English court position, as expounded in *Wah v. Grant Thoronton International Ltd* [2013] 1 Lloyd's Rep 11, is that 'agreement to negotiate in good faith, without more, must be taken to be unenforceable'. 'Good faith' cannot be tangibly defined. Therefore, in this case, the judge set out several conditions to be met for a mediation clause and they include: (a) a certain and unequivocal commitment to commence a process; (b) the steps that each party needed to take to put a process in place; (c) the minimum level of participation of both parties in the process; and (d) when or how the process would be exhausted or capable of being terminated without breach.

Pragmatically, cost consideration is a genuine concern because preparing for mediation involves disproportionate costs in comparison to the modest value of claim. Moreover, when the opponent was viewed as 'intransigent', mediation was thought to be fruitless. A common thread running through many of the 'inappropriate' objections was a sense that for mediation to have any prospect of success there must be scope for compromise and some willingness on either side to move towards such a compromise. In the absence of such willingness, automatic referral to mediation was an unwelcome hurdle in the course of the litigation. Furthermore, there are certain prerequisites for a successful mediation. If parties attended a mediation with a willing spirit to negotiate and compromise, then the chances of achieving a settlement, even within the three-hour limit, appeared to be good. The skills of the mediator are also pivotal. Skillful mediators who make pre-appointment contact; read the papers and show that they are familiar with the issues; know the law; take a proactive stance; and practise effective shuttle-diplomacy are more likely to gain a settlement. The prospect of a mediation hinges on the ability of the parties to attain fruitful exchanges and understand the strengths and weaknesses of cases.

What attracts the advocates of mandatory mediation is the claimed efficiency or utility that could be brought about by mediation, including cost and time advantages, and court resource savings. Moreover, there have been genuine concerns that compulsion of mediation can lower the admirable settlement rate of voluntary mediation. A failed mandatory mediation will even add time and costs consumed to resolve the dispute, so without a higher success rate it is unattractive to make mediation mandatory, even though time and cost savings are achieved. Despite such doubts, it is generally believed that mandatory mediation can indeed improve the efficiency of resolving disputes. But is such efficiency justified in sacrificing some merits of voluntary mediation? Meggit (2012) expressed his concern that 'voluntariness and party empowerment are only justified when they lead to mediation and if they do not, they should be sacrificed in the name of efficiency'. The Working Party suggested that 'it is in the interests of justice to promote cost-effective options for a fair and satisfactory dispute resolution' in the final report, which means it is believed that court-connected mediation is able to achieve the same result (a fair and satisfactory resolution) as going to court, but in a potentially more cost-effective manner. The drive for utility can be illustrated by highlighting the attributes of mediation that are by nature utility-driven (Table 4.5 refers to this and utility is indicated by 'U').

However, and as a matter of fact, the settlement agreement is very different from the court decision in that a mediation settlement is not made based on any legal rules but based on a distribution of value which both sides could accept (David, 2004). A settlement agreement involves too many factors that will not be

Table 4.5 Attributes of mediation

Attributes of mediation	Characteristics	Attributes of mediation	Characteristics
Confidentiality (U)	Yes	Power to compel consolidation	No
Choice of adjudicator or appointee (U)	Yes	Width of remedies (U)	Not applicable
Range of issues (U)	Open ended	Binding decision	No
Flexibility of procedure (U)	Very high	Enforceability of decision	Not applicable
Delay potential (U)	Very low	Susceptibility to appeal	Not applicable
Control by parties (U)	Very high	Liability for opponent's costs (U)	None
Susceptibility to tactics (U)	Very low	Cost of tribunal (U)	Parties as agreed
Control over parties	None	Level of costliness (U)	Low
Control over witnesses	Not applicable		

Source: Adapted from Cheung, 1999.

considered in a court litigation, such as previous relationships, the characteristics of disputants, emotional factors, negotiating skills of the disputants, and so on. Therefore, each mediation settlement is special and won't really have any reference value for future similar cases. On the contrary, judges basing decisions on laws and facts decide litigation cases. Judgments are not made for the purpose of settlement but should be based on legal principles. Decided cases are of great value for future cases as precedents. On this note, Roebuck (2011) made the point that mandatory mediation may threaten legal development as mandatory mediation may reduce the flow of litigation to the level where the development of new law is endangered. To establish norms, decisions must be regularly recorded and published and norms cannot be established by private contract made in mediation (Roebuck, 2011). Lord Neuberger (2011) expressed that:

> unless there is a healthy justice system, with judges developing law to keep pace with the ever accelerating changes in social, commercial, communicative, technological, scientific and political trends, neither citizens nor lawyers will know what the law is. Imperfect or outdated law will harm mediation and other alternative dispute resolution approaches as well, because without the law supporting the party's stance, the party will be in a weaker position to mediate as well.

By court-connected mediation, the development of law in certain disfavored areas of cases might be stifled. Edwards (2015) used an example that if all race discrimination cases in the sixties and seventies had been mediated rather than going to litigation, the wholesale diversion of cases involving the legal rights of the poor may have resulted in the definition of these rights by the powerful in our society, rather than by the application of fundamental societal values reflected in the rule of law. The diversion of cutting-edge cases from the formal justice system might lead to so-called 'second-class justice' and another example used is that women have gained many new rights through litigation and if all 'family law' disputes are blindly mediated (mediation of a dispute without due consideration of other options), these new rights may become nothing but a mirage. Especially when the disputes pursuant to an ADR system involve significant public rights and duties, court-connected mediation may result in an abandonment of our constitutional system (Edwards, 2015). Edwards (ibid.) also stated that if ADR is extended to resolve issues of constitutional or public law, it is likely that non-legal values may be applied to resolve important social issues and the law may delimit the legal rights of people. Fiss (1984) states that the job of officials in adjudication:

> is not to maximize the ends of private party or to secure the peace or relationship between parties, not simply to secure peace, but to explicate and give force to the values embodied in authoritative texts such as Constitutions or Statutes: to interpret those values and to bring reality in accord with them.
> (Fiss, 1984)

J. Anthony Lukas (2015) highlights that in many occasions and disputes, the most fundamental values in the society reflected in rule of law, such as equality and access to justice under the law is in conflict with local non-legal ends embraced in ADR approaches. It can therefore be seen that the major concern over the use of mediation from the legal profession is the shifting emphasis towards utility instead of observing the rule of law.

In addition, there are concerns that ADR will gradually replace the courts and litigation. Lord Neuberger (2011) expressed his view in the 'Swindlers Not Wanted' lecture (2011) and warned that the alternatives cannot be the norm, and the court's role of protecting rights cannot be ignored. He also expressed his concern on the access of all to the courts in the lecture:

> In our modern consumer, market-based society, with its multiplicity of laws and rights, and its increasing scope for legal disputes, it is more important than ever that we have effective, accessible institution of law. If not, laws go unenforced. They cease to be rights, but rather become privileges for those select few who can afford them … the irreducible cost of a genuinely accessible and truly effective legal system has to be paid if we wish to remain a civil society.

Edwards (2015) commented that before embracing any form of ADR, it is imperative to consider whether the proposed ADR is to facilitate existing court procedures or an alternative for the existing court procedure. He expressed his concern on ADR 'becoming a tool for diminishing the judicial development of legal rights for the disadvantaged'. He quoted Professor Tony Amsterdam's words delivered in the 1984 in Judicial Conference that 'in the name of increased access to justice and judicial efficiency', fair and just legal redress of wrongs that the poor are indeed suffering from can be impaired, especially where the two parties do not possess equal bargaining power. In spite of its advantages regarding costs, time and informality, ADR could lead to 'ill-informed decisions'. Edwards (2015) also commented that compulsory mediation could make litigation for dispute resolution deviant. Tully (2009) expressed his worry that disputants may be considered 'components of a problem' and their original rights to access the courts could be neglected.

'Good faith participation' is another concern widely raised. Currently, Hong Kong uses the construct of 'minimum participation level' to objectively regulate the time or sessions to be attended by the parties. But with court-connected mediation, whether such requirements will be changed to 'reasonable attempts' or 'bona fides' is attracting attention. The main drawback of 'minimum level of participation' is that where one party is genuinely unwilling to settle, the requirement to attend a minimum number of sessions does not improve the prospect of having a settlement outcome, but merely to increase the costs incurred in the process. The problem brought by 'good faith participation' is that the parties may be under pressure to settle on an unattractive and dissatisfactory offer in fear of the allegation of 'bad faith participation'. As a result, the paradox of satellite litigation will appear where more court time is to be used to deal with this process that is originally designated to save time. Meggit (2008) also questioned whether 'good faith participation' would impede the confidentiality of mediation on the ground that a mediator can report or seek to argue a 'lack of good faith' in court. Such an occurrence would weaken trust and affect the purpose of 'open talk'.

Furthermore, the utility side of ADR has not been fully addressed because for disputants who are forced to use mediation but haven't settled their cases in mediation, they will need to bear with even more costs and further time delays. Whether mandatory mediation will have a lower settlement rate than voluntary mediation remains unsettled (Genn et al., 2007). If court-connected mediation were applied, whether 'good faith participation' or 'minimum participation level' is adopted would be critical, as the former may lead to breach of confidentiality and accusation of coercion to settle (Meggit, 2008).

Notwithstanding, Leung (2014) highlight cases not suitable for mediation. These include: (i) policy at stake affecting public interest; (ii) dispute over a legal point; (iii) parties not interested or with bad faith; (iv) cases involving a risk or threat of personal danger; (v) issues on child abuse, family violence or criminal activities; (vi) parties involved are too emotional; (vii) there is no room for compromise; (viii) remedies that only a court can provide and (ix) there is a serious imbalance of power. There are pros and cons of having court-connected mediation. The bottom line appears to be that in no way must the rule of law be undermined. Distinguishing cases for the suitability of court-connected mediation is no easy task.

Summary and conclusion

One of the aims of using alternative dispute resolution to resolve construction disputes is to reduce the number of cases reaching the court system. Even if a case is due for hearing, in some jurisdictions court-connected mediation has been applied to enable a last attempt to get the dispute settled by the parties themselves. In Hong Kong, at project level, contractual use of mediation is adopted. The use of mediation within the court system reached a highpoint when the Civil Justice Reform came into effect in 2009. After several pilot schemes conducted to test the outcomes, the Hong Kong judiciary promulgated Practice Direction 6.1 under which an 'indirect' mandatory mediation is now applied to cases on the Construction and Arbitration List. PD 6.1 enables the court to give adverse cost orders to the party who unreasonably refuses or fails to mediate to the minimum participation level when there is a Mediation Notice. With reference to other jurisdictions, especially in Australia, the United Kingdom, and Singapore, it is interesting to explore whether a direct court-connected mediation could be implemented for construction disputes in Hong Kong. The advantages and benefits claimed by such an approach have been widely recognized by advocates. However, there are concerns that the benefits of voluntary mediation, such as high settlement rates, cannot be maintained with mandatory mediation. Without a higher success rate, court-connected mediation can only lead to higher costs and more waiting time at the end because the disputants will need to bear both the mediation and litigation costs. The legal profession is critical and has expressed the need to rethink the over-riding objective of ADR as a dispute resolution mechanism. Does this suggest that there is an inherent tension between the utilarian nature of ADR and the need to uphold the rule of law? It could be argued that disputants who insist on litigation should not be punished for protecting their rights by having disputes decided in court. It is also claimed that with more and more cases going to mediation rather than litigation, rights and public interests that should be recognized and protected by the court's decisions will only become issues in private discussions and settlements. With these concerns, Hong Kong has not installed court-connected mediation. The experience in the use of voluntary mediation under PD6.1 and PD31 will inform further development in this subject matter.

Acknowledgements

Special thanks is given to Miss Lucia Lu Shen and Miss Mavis Ho Ching Cheung for collecting data for this study.

Notes

1 Supreme Court Mediation of NSW. www.Supremecourt.justice.nsw.gov/supremecourt/sco2_mediationinthesc/court_connected_mediation.html.
2 The rate calculated by only taking matter types that are more commonly referred to mediation into account and excluding those whose characteristics mean that referrals to mediation are rare such as migration appeals.

References

Amsterdam, A.G. (1984), 'Too many lawyers, too many suits, not enough justice'. Forty-Fifth Judicial Conference, D.C. Circuit, USA.

Boulle, L., and Rycroft, A. (1998), *Mediation: Principles, Process, Practice*, Durban: Butterworth.

Brunsdon-Tully, M. (2009), 'There is an ADR but does anyone know what it means anymore?', *Civil Justice Quarterly*, 28(2), 218–237.

Census and Statistics Department, Government of The Hong Kong Special Administrative Region (CSD), (n.d) *Hong Kong Statistics*, available online at: www.censtatd.gov.hk [Accessed 13 June 2015].

Cheung, S.O. (2009), 'Construction mediation in Hong Kong', in S. Wilkinson and P. Brooker (eds). *Mediation in The Construction Industry: An International Review*, London: Spon Press.

Cheung S.O. (2014) *Construction Dispute Research: Conceptualisation, Avoidance and Resolution.* New York: Springer.

Cheung, S. O. (1999), 'Critical factors affecting the use of alternative dispute resolution processes in construction', *The International Journal of Project Management*, 17(3), 189–194.

Davis, M., Davis, G. and Webb, J. (1996), *Promoting Mediation: Report of a Study of Bristol Law Society's Mediation Scheme in its Preliminary Phase*, Research Study no. 21, London: The Law Society.

Department of Justice of HKSAR (2004), *Chief Justice's Working Party on Civil Justice Reform, Civil Justice Reform Final Report*. Hong Kong: Chief Justice's Working Party on Civil Justice Reform.

Edwards, H.T. (1986), 'Alternative dispute resolution: panacea or anathema', *Harvard Law Review*, 99(3), 668–684.

Feinberg, K.R. (1991), 'Using mediation to resolve construction disputes'. In: R.F. Cushman, G. C. Hedemann and A.S. Tucke, (eds) *Alternative Dispute Resolution in the Construction Industry*, Chichester: Wiley Law Publications, 192–210.

Fiss, O. M. (1983), 'Against settlement', *Yale Law Journal*, 93, 1093.

Genn, D.H., Fenn P., Mason M., Lane A., Bachai N., Gray L. and Vencappa D. (2007), *Twisting Arms: Court Referred and Court Linked Mediation under Judicial Pressure*, Ministry of Justice Research Series, 1/07. London: Ministry of Justice.

Gillie, P., Goetz, P., Muller, F., and Feinberg, K.R. (1991). 'Using mediation to resolve construction disputes'. In: R.F. Cushman, G.C. Hedemann and A.S. Tucke, (eds) *Alternative Dispute Resolution in the Construction Industry*, Chichester: Wiley Law Publications, 153–210.

Hong Kong International Arbitration Centre (2008–2013). *Case Statistics*. Available online at: www.hkiac.org/en/hkiac/statistics. [Accessed 13 June 2015].

Hann, R.G., Baar C., Axon L., Binnie S. and Zemans F. (2001), *Evaluation of the Ontario Mandatory Mediation Program, Rule 24.10 Final Report – The First 23 Months*, Ottawa: Civil Rules Committee, Evaluation Committee for the Mandatory Mediation Pilot Project.

Hilmer, S.E. (2012), 'Mandatory mediation in Hong Kong: a workable solution based on Australian experiences', *China–EU Law Journal*, 1:61–96.

Ingleby, R (1993), 'Court sponsored mediation: the case against mandatory participation', *The Modern Law Review Limited*, 56:3.

Lees, A (2015), 'The enforceability of negotiation and mediation clauses in Hong Kong and Singapore', *Asian Dispute Review,* January, 16–21.

Leung R. (2014) '*Hong Kong Mediation Handbook*', London: Sweet & Maxwell.

Lukas, J.A. (1985), *Common Ground: A Turbulent Decade in the Lives of Three American Families*. New York: Vintage.

Meggitt, G (2008), 'The case for (and against) compulsory court-connected mediation in Hong Kong', 5th Asian Law Institute Conference, 22–23 May, Singapore.

Nesic, M. (2001), 'Mediation: on the rise in the United Kingdom'. *Bond Law Review*, 13(2), Article 10. Available at: http://epublication.bond.edu.au/blr/vol13/iss2/10. [Accessed 30 Spetember 2016].

Neuberger D. (2011), 'Swindlers (including the Master of the Rolls?) not wanted', Bentham and Justice Reform, Bentham Lecture delivered at University College, London, March.

Onn, L.S. and Koh, D., (2015). 'The Singapore Mediation Centre (SMC)'. In D. McFadden and S.G. Lim (eds) *Mediation in Singapore – A Practical Guide*. Singapore: Sweet & Maxwell, p. 248.

Prakash, J. J., 2009. 'Making the civil litigation system more efficient'. Available online at: www.apjrf.com/papers/Prakash_paper.pdf [Accessed 29 June 2015].

Quek, D., 2015. 'The development of mediation for civil disputes'. In D. McFadden and S. George Lim, (eds) *Mediation in Singapore – A Practical Guide*. Singapore: Sweet & Maxwell, p. 272.

Roebuck, D (2011), 'Time to think: understanding dispute management', *Arbitration*, 77(3), 342–350.

Zampano, H.R.C. (1991), 'Court-connected ADR: A view from the bench', In R.F. Cushman, G.C. Hedemann and A.S. Tucke, (eds) *Alternative Dispute Resolution in the Construction Industry*, Chichester: Wiley Law Publications, 211–235.

List of cases

Golden Eagle International (Group) Limited and Gr Investment Holdings Limited HCA 2032/2007

Pacific Long Distance Telephone and New World Telecommunications Limited HCA 1688/2006

Supply Chain & Logistics Technology Limited and NEC Hong Kong Limited HCA 1939/2006

Halsey v. Milton Keynes General NHS Trust [2004] EWCA Civ. 576

McCook v. Lobo [2002] EWCA Civ. 1760

Burchell v. Bullard [2005] EWCA Civ. 358

Walford v. Miles [1992] 2 AC 128

Channel Tunnel Group Ltd v. Balfour Beatty Construction Ltd [1993] AC 334

Kenon Engineering Ltd v. Nippon Kokan Koji Kaisha [2003] HKCFI 568

Hyundai Engineering & Construction Co Ltd v. Vigour Ltd [2004] HKCFI 205

Wah v. Grant Thoronton International Ltd [2013] 1 Lloyd's Rep 11

List of Statutes Practice Direction 31, Hong Kong High Court

Practice Direction 6.1, Hong Kong High Court

Hong Kong Arbitration Ordinance (Cap 609)

Hong Kong Legal Aid Ordinance (Cap 91)

Hong Kong Mediation Ordinance (Cap 620)

Supreme Court Amendment (Referral of Proceedings) Act 2000

Civil Procedure Act 2005 (NSW, Australia)
Victorian Civil and Administrative Tribunal (VCAT, Australia)
Magistrate's Court Act 1989 (VIC, Australia)
Civil Procedure Act 2010 (VIC, Australia)
Federal Court of Australia Act 1976 (Australia)
Law and Justice Legislation Amendment Act 1997 (Australia)
Courts (Mediation and Arbitration) Act 1991(Australia)
Family Law Rules 2004 (Australia)
Administrative Appeals Tribunal Act 1993 (Australia)
Women's Charter (Cap 353) (Singapore)
Protection from Harassment Act 2014 (Singapore)
Supreme Court of Judicature Act 2014 (Singapore)

5 Court-connected mediation for construction disputes in the United States of America

Laura J. Stipanowich

Introduction

The construction industry played a major role in the development of alternative dispute resolution (ADR) processes in the United States of America ('US'). The vast majority of modern US construction contracts require parties to engage in 'ADR' and, until recently, mandated arbitration, not litigation, as the ultimate means of dispute resolution. Moreover, most US courts, both on the federal and state level, have implemented their own ADR programs.

As of mid-2015, following a decade-long downturn in the economy, US construction spending is increasing. This is demonstrated in monthly reports published by the United States Census Bureau of the Department of Commerce:

Table 5.1 Construction spending (seasonally adjusted, in millions of dollars)

Year	Month	Total	Private construction	Public construction
2005	July	1,088,324	856,158	243,165
2006	July	1,199,990	930,928	269,062
2007	July	1,169,074	880,070	289,004
2008	July	1,084,355	774,635	309,720
2009	July	958,037	630,422	327,615
2010	July	805,159	506,359	298,800
2011	July	789,510	514,499	275,012
2012	July	834,384	558,714	275,670
2013	July	900,824	631,403	269,422
2014	July	981,305	701,657	279,648
2015	July	1,083,378	787,776	295,602

Source: United States Census Bureau, Construction Spending.

Along with increases in spending come increases in the number of disputes. A party considering a construction project in the US must be educated about ADR options for construction dispute resolution – both informal and formal – to craft an agreement that meets the unique needs of the project and the parties involved.

Federal and state judicial and administrative systems now incorporate ADR methods to supplement the traditional litigation model. A comprehensive study of each court system's ADR program is not possible within this chapter. As one expert astutely noted, 'the nuanced variations in ADR programs, and the technological balkanization of the courts have combined to largely prohibit a systematic analysis of the many state and federal programs, even as a descriptive matter.'[1]

Given these constraints, this chapter shall focus generally on court-connected ADR and its ramifications for the US construction industry. First is a brief overview of the forms of private ADR common to the US construction industry. Next is an exploration of the US court system, with a focus on current court-connected ADR programs in the federal and state courts, respectively. Lastly, this chapter shall look briefly at the commentary on the relative merits and perceived weaknesses of court-connected ADR in the construction context.

Private alternative dispute resolution in the US construction industry

The construction industry was one of the driving forces behind the inception of US ADR. Construction disputes are often too complex to make traditional litigation a cost-effective option and involve parties who may have to work together on future projects, prioritizing conciliatory efforts to resolve conflict. Construction disputes typically are resolved prior to trial, either through informal dispute resolution processes like mediation, non-binding arbitration, or court-connected ADR programs. If the parties fail to resolve their dispute, they will head to court litigation or binding arbitration.

Mediation

In the US, mediation proceedings are, with rare exceptions, kept private and confidential, which is key to their popularity. This mediation privilege has been the focus of a number of federal and state cases.[2] The non-binding nature of mediation is also critical to its popularity. Unlike arbitration, parties to mediation control its ultimate resolution and do not risk an adverse award.

The National Conference of Commissioners on Uniform State Laws (NCCUSL) has calculated that 'legal rules affecting mediation can be found in more than 2,500 statutes.'[3] This can lead to conflict and inconsistency. In an effort to clarify and systemize mediation throughout the US, the National Conference of Commissioners on Uniform State Laws approved the Uniform Mediation Act ('UMA') in August 2001 and the American Bar Association approved it on February 4, 2002.[4]

The UMA generally applies to mediation in which: (1) the parties are required to mediate by statute or court or administrative agency rule or referred to mediation by a court, administrative agency, or arbitrator; (2) the parties and mediator agree to mediate in a record that demonstrates an expectation that mediation communications will be privileged against disclosure; or (3) the parties use a mediator who holds himself out as a mediator, or mediation conducted by a mediation service.[5] The primary function of the UMA is to clarify the unique, confidential nature of mediation. UMA §§ 3–9 outline standardized privilege and confidentiality requirements (and exclusions thereto) in the mediation process, and § 10 governs mediator disclosure requirements.

As of this writing, the District of Columbia and 11 out of the 50 US states have enacted the UMA as written and the legislatures of both Massachusetts and New York have introduced the UMA for potential adoption.[6] Many states that have not enacted the UMA, however, have their own mediation acts dictating privilege, confidentiality, and disclosure.

Arbitration

Arbitration has long been the preeminent form of ADR in the US construction industry. In 1915, before the arbitration process itself was sanctioned by statute, an arbitration clause was included in the American Institute of Architects' (AIA) Standard General Conditions.[7] The American Bar Association (ABA) in 1921 developed a draft of a Federal Arbitration Act ('FAA'), codified at 9 USC. § 1, *et seq.*, which became law in 1925. Facing a judiciary reluctant to embrace ADR, the FAA compelled federal courts to recognize that arbitration agreements were 'valid, irrevocable, and enforceable …'[8]

After initial reluctance, the US federal courts – in particular, the US Supreme Court – had 'transformed the FAA from a mere procedural statute regulating the practice of arbitration in the federal courts into a substantive law of arbitration having wide ranging impact on contracts involving interstate commerce containing arbitration agreements.'[9] With limited exception, the FAA applies to arbitration of disputes affected by interstate commerce – practically speaking, almost all construction projects.[10]

The mid-twentieth century also saw the development of state arbitration statutes, spurred on by the establishment of the Uniform Arbitration Act (UAA) in 1955 and the Revised Uniform Arbitration Act in 2000 (RUAA).[11] The RUAA, inter alia, provides for consolidation of arbitrations involving multiple parties, requires arbitrators to disclose potential conflicts of interest, and provides for the arbitrator's judicial immunity. Currently, almost all 50 states have adopted arbitration statutes.[12] These statutes often resemble the FAA and state courts may rely upon federal case law interpreting the FAA when construing state arbitration statutes.

In addition to statutory and judicial efforts to further ADR, non-profits like the American Arbitration Association (AAA)[13] and the International Institute for Conflict Prevention and Resolution ('CPR'),[14] as well as for-profit organizations like the Judicial Arbitration and Mediation Services ('JAMS'),[15] have played a role in furthering US arbitration. These organizations develop uniform rules, including industry-specific rules like the AAA's 'Construction Industry Arbitration

Rules and Mediation Procedures' and CPR's 'Rules for Expedited Arbitration of Construction Disputes,' and maintain lists of trained mediators and arbitrators. These procedures – even in complex disputes – are more streamlined than US litigation, with clear protocols and deadlines.

The AAA rules, for example, mandate a fee structure and step-by-step rules for construction mediation/arbitration – including separate dispute procedures delineated as 'regular,' 'complex' (claims involving multiple parties totaling more than $1,000,000) and 'fast track' (claims between two parties totaling under $100,000).[16] A typical AAA arbitration begins with the filing of a demand, followed by answers and counterclaims. The parties will then mutually agree upon the locale for the arbitration and will choose an arbitrator or panel (if they are unable to agree, the AAA will dictate the locale and assign an arbitrator). The AAA arbitration then proceeds, consisting of, inter alia, preliminary hearings, presentation of evidence, filing of dispositive motions, and, usually, oral argument(s).[17] Ultimately, the arbitrator or panel will render an award, typically no later than 30 days from the closing of the hearing/submittal of final statements.

Although arbitration remains a popular forum for adjudication of construction disputes, the last decade has seen a gradual shift in preference from private arbitration to traditional litigation. For example, until 2007 the AIA, which provides contract forms utilized by many US construction participants, included arbitration as the default dispute resolution mechanism in its contracts, but now defaults to traditional litigation.[18]

A major factor behind this is the lack of appellate recourse in arbitration. While a dissatisfied litigant may appeal, the losing party in an arbitration has, effectively, no recourse. In only the most limited of circumstances will a court of law quash an arbitration award and the Supreme Court has repeatedly condemned court interference with the arbitration process.[19] This has led many construction participants, especially those that have been on the losing end of an arbitration, to retreat from their prior preference for arbitration.

Other forms of ADR

Although mediation and arbitration remain the most popular forms of ADR in the US construction industry, they are not the only options available. Although a detailed discussion of these additional ADR forms is impractical for the purposes of this chapter, they generally include the following (some of which are implemented by the court-connected ADR programs):

Project partnering

'Project partnering' in the US is 'a process by which multiple parties work collaboratively before and during a project to develop and implement systems, procedures, mechanisms, metrics, incentives – whatever the alliance needs to achieve shared success and value for all participants.'[20] This approach 'formalizes the following aspects of enterprise planning traditionally employed by successful contracting entities: getting to know your contracting partner; identifying common goals and discussing specific plans and expectations; and establishing clear channels of communication and fail-safe mechanisms for resolving potential problems.'[21]

Despite initial enthusiasm for the concept of partnering, commentators have noted that 'expecting the parties to change their behavior with nothing more than a partnering agreement to support them,' is naive.[22] Partnering, however, does continue to be a valuable tool for dispute avoidance, especially if utilized in conjunction with other ADR methodologies.

Dispute Review Boards, Project Neutrals, Initial Decision Makers

The 'Dispute Review Board' (DRB) and 'Project Neutral' or 'Initial Decision Maker' (IDM), commonly referred to as 'Early Neutral Evaluation' ('ENE'), allows parties an opportunity to negotiate without resorting to formal ADR or litigation. DRBs can be found on 'significant construction projects such as heavy civil, wastewater, medical, and manufacturing plants, as well as airport, power plant, sports complexes and office building construction, among others.'[23] The DRB or Project Neutral will regularly visit the project site in order to issue non-binding recommendations for prompt dispute resolution. In the US, on-site dispute resolution methods have been used for over 100 years and continue to grow in popularity.[24] An increase in the utilization of DRBs and Project Neutral/IDMs is evident in the standard form agreements being used on modern construction projects. For example, reliance on a project neutral is one of the dispute resolution options recognized by the 2007 ConsensusDOCS (Article 12 of ConsensusDOCS 200) and the 2007 AIA (A201) contract forms.[25]

Structured negotiation

Many project-level disputes are not resolved due to the paucity of information available to the parties. To combat this, many US construction contracts include a stepped ADR clause, a 'structured negotiation' process. Commentators describe this ADR method as a formal procedure (1) for full disclosure of information, (2) for timely commencement of project-level negotiations, and, if needed, (3) for moving negotiation up to in the parties' respective organizations' higher management before seeking a third-party non-binding recommendation or binding decision.[26]

Expert determination

Parties reluctant to engage a standing neutral, IDM or DRB, may find it beneficial to retain an expert to make determinations on an as needed basis.[27] This expert determination is comparable to court appointment of experts and special masters under Federal Rule of Evidence 706 and Federal Rule of Civil Procedure 53, which allow an expert to hold trial proceedings and render findings of fact and conclusions of law on issues to be decided by the court without a jury.[28]

The US legal system

Sources of law in the United States

The United States has a federal system of government. The US Constitution is the ultimate legal authority, which governs all 50 states. Any federal or state law in conflict with the US Constitution may be overruled and unenforceable. Subordinate to the US Constitution are the individual state constitutions.

In addition to constitutional law, the US has a complex system of statutory laws and rules enacted by the US Congress, and state and local legislative bodies. Procedurally, federal law governs federal court, while state, municipal and local laws govern state courts. Substantively, federal law will apply to resolve construction disputes involving the United States Government, while the law of one of the 50 states will apply to resolve all other domestic construction disputes.

A great deal of US law is based on the English common law tradition, wherein trial courts follow established precedent in rendering decisions to maintain consistency. This is the practice of *stare decisis*, literally 'to stand by things decided,' whereby courts rely on prior precedent when determining how to decide a particular matter. Prudent parties consider whether a jurisdiction's common law precedents are favorable when determining whether to proceed in public litigation or private ADR.

Lastly, construction projects are subject to laws created by administrative agencies, which have specific authority to make and enforce rules relating to their stated purpose. For example, the Environmental Protection Agency can write and enforce regulations to protect human health and the environment, like the Clean Air Act.[29]

The US court system

There are two primary court systems in the United States: federal and state. The forum for a case is determined by the jurisdiction of the court. 'Jurisdiction' is the practical authority to interpret and apply the law, by which courts and judicial officers take cognizance of and decide cases.[30]

In the federal and state courts, generally, there are two classes of cases: criminal and civil. Most construction disputes are civil disputes, in which a party seeks redress or compensation from another. In the context of construction law, civil cases are used as a means to resolve disputes involving damages sounding in contract, tort, and equity.

Federal courts

The Constitution only allows certain kinds of cases to be heard by the federal courts. In general, federal courts have jurisdiction over the following categories of cases:

* Constitutional law and disputes;
* Matters affecting interstate commerce or relations;
* International law and relations; and
* Disputes involving both federal and state law.

The US federal court system is tiered and divided into separate districts, appellate circuits and one Supreme Court. All 50 states, the District of Columbia, Guam, Puerto Rico, and the US Virgin Islands have a federal District Court, and some have multiple districts depending on size and population. There are 11 federal circuits, each of which encompasses more than one federal district and each is home to a Federal Court of Appeal.

Certain types of construction disputes are heard in special courts, such as the Court of Federal Claims, which has jurisdiction 'to render judgement upon any claim against the United States founded either upon the Constitution, or any Act of Congress or any regulation of an executive department, or upon any express or implied contract with the United States, or for liquidated or unliquidated damages in cases not sounding in tort.'[31]

The US Supreme Court sits atop the federal judicial hierarchy. To qualify for an oral argument before the Supreme Court, a case must pass through the state and/ or federal system, and the litigants must successfully file a petition for a writ of certiorari. On average, the Supreme Court only grants hearings to 75–80 petitions out of the approximately 10,000 submitted per year.[32]

State courts

Matters pertaining to a single state or arising under a state's local laws will be heard in state court. Most US cases are filed in state court. Most states have trial courts of *limited jurisdiction* and of *general jurisdiction.* Courts of limited jurisdiction include municipal and magistrate courts, which hear minor criminal cases and civil lawsuits. Courts of general jurisdiction, which hear cases involving higher damages and monetary claims, may include circuit courts, superior courts, district courts, or courts of common pleas. Many states also have specialized trial courts, such as probate and small claims courts.

Most states have appellate courts, which serve a function analogous to that of the Federal Circuit courts. Parties who are unsatisfied with the findings of a trial court may seek relief from the appellate court. Appellate courts do not hold trials, but have the authority to uphold, amend, or reverse trial court decisions, or to remand cases back to the trial courts for further disposition.

Finally, states have their own supreme courts, which sit in review of lower courts' decisions and are the courts of last resort. In limited circumstances, State Supreme Court decision may be appealed to the US Supreme Court.

Court-connected alternative dispute resolution in the United States

Formalized ADR programs have been available in US courts since the 1970s. Many distinctions exist between private, court-connected, and court-connected ADR.[33] Therefore, some independent agencies have developed model standards for court-connected ADR, which have been adopted (in whole or in part) by US courts. For example, the Center for Dispute Settlement and the Institute of Judicial Administration jointly developed the National Standards for Court-

Connected Mediation Programs (NSCCMP) to 'guide and inform courts interested in initiating, expanding or improving mediation programs to which they refer cases.'[34] The NSCCMP apply to all court-connected programs, and cover issues including availability of mediation services, education of pro se litigants about mediation, providing mediation regardless of a party's ability to pay, and educating parties and their attorneys about the program(s).

The NSCCMP also stresses that, if there is a shortage of resources, decisions about referrals to mediation should be made based upon clearly defined parameters.[35] The NSCCMP mandate courts' responsibility to monitor ADR programs' performance.[36] Mandatory referral to mediation should come only after an assessment of whether mediation is appropriate.[37] Mandatory attendance at an initial mediation session, for example, may be appropriate, but only when such a requirement is more likely to serve the interests of parties, the justice system, and the public than would voluntary attendance.[38] The NSCCMP also outline the court's responsibility for mediators (1) employed by the court, (2) who receive referrals from the court, or (3) who are chosen by the parties; and encourage adoption of a code of ethics for mediators.[39]

Procedurally, federal and state courts approach court-ordered mediation in similar fashion. Each court will, of course, have some unique protocols, but experienced construction practitioners are well-versed in the procedures of their local courts. Generally, the court will order the parties, often by way of a case management order, to complete mediation or any other court-connected ADR procedure by a date certain. The parties then have leeway to procure an independent neutral, if permitted, or retain a neutral from the court's roster.

Parties prepare for a court-connected mediation in much the same way they would do so for a private mediation. Parties consult with counsel and put together a Settlement or Mediation Statement laying out their positions for pre-mediation submission to the neutral. If time and the forum allows, they assemble presentations (often a PowerPoint) to lead off the mediation session. Like a private mediation, the parties and mediator control the procedural aspects of the court-ordered mediation, such as whether parties will submit mediation statements, conduct individual caucuses, or make opening presentations.

Upon conclusion of the ADR process, the parties (or court-appointed neutral) will file a short document with the court attesting that the ADR requirement was met and indicating whether settlement was achieved. Just as with a private mediation, the topics discussed and any settlement discussions remain strictly confidential.

Some courts, like those in the State of Maryland, have ADR procedures that are somewhat more formalized. In Maryland state court, the parties are ordered to a pretrial settlement conference, at which both counsel and their clients must appear, and meet with a neutral (often a retired judge). Prior to the settlement conference, each party will submit a statement summarizing their respective positions. The meeting proceeds much like a private mediation, with the neutral determining whether to caucus individually with the parties and trying to facilitate an amicable resolution, following which the neutral will report back to the court regarding the outcome.

ADR in the federal courts

Beginning in the 1970s, federal courts began to experiment with ADR to reduce caseloads. In 1988, following test runs of court-connected arbitration by the Judicial Conference, Congress enacted the 'Judicial Improvements and Access to Justice Act,' authorizing a subset of district courts to mandate arbitration for specific civil cases and to offer voluntary use of arbitration.[40]

Following complaints about delays in US federal courts, Congress implemented the 'Civil Justice Reform Act' of 1990 (28 USC. §§ 471–482) (the 'CJRA'). The CJRA authorized 13 district courts to adopt ADR programs and required other district courts to 'experiment with various methods of reducing cost and delay in civil litigation, including alternative dispute resolution.'[41] The CJRA stated this 'may be a plan developed by such District Court or a model plan developed by the Judicial Conference of the United States,' and suggested six case management principles, including one for alternative dispute resolution.[42]

In response to the CJRA, most of the 94 district courts developed ADR programs and some hired professional staff to manage them. The CJRA offered monetary incentives for achieving objectives. In 1997, as required by the CJRA, the Court Administration and Case Management Committee (CACM) submitted a report to Congress summarizing the district courts' implementation of the CJRA. They recognized that '[m]any courts have shown the ability and commitment to administering court-connected ADR programs' and recommended that 'local districts continue to develop suitable ADR programs.'[43]

After the expiration of the CJRA,[44] Congress passed the Alternative Dispute Resolution Act of 1998 (28 USC. §§ 651–658) ('ADRA'), which mandates inter alia that district courts create ADR programs. The adoption of court-connected arbitration in the federal courts has arisen largely through the ADRA, which allows federal courts to establish mandatory ADR programs and mandates the establishment of voluntary ADR programs. Although many districts had adopted ADR programs under the CJRA, some improved on existing programs, and a number of districts that had not previously adopted ADR programs authorized them (without creating specific ADR programs).[45]

In 2011, the Federal Judicial Center conducted a preliminary study of the programs implemented and prepared an Initial Report ('FJC Report') codifying the data collected.[46] The report contained tables, reproduced herein below, that provide information about 'the types of ADR programs adopted by the district courts, the procedures by which cases are referred to ADR, the providers of ADR services (i.e., the neutrals), the fees paid to ADR neutrals, and the number of cases referred to ADR.'[47] Although the report is a few years old, it provides an excellent snapshot of ADR in the federal system.

As Table 5.2 demonstrates, over a third of the courts polled by the FJC in 2011 had multiple ADR options for litigants or a 'stepped' procedure, while another third rely upon one option only. The FJC Report noted that, typically, districts with multiple forms of ADR authorize either: mediation and arbitration, mediation and ENE, or a combination of all three.

Table 5.2 Breakdown of ADR procedures in federal courts

Type of ADR Procedures authorized	Number and percentage of district courts	
	Number	Percent
Multiple forms of ADR[a]	34	36.2
Mediation Only[b]	27	28.7
General authorization only	12	12.8
Settlement conference only	10	10.6
General authorization[c] and settlement conference only	3	3.2
Early neutral evaluation (ENE) only	3	3.2
Summary jury trial only	1	1.1
Case evaluation[d] only	1	1.1
Other[e]	3	3.2
Total	94	100.1

Notes:

a. A district is counted in this category only if it authorizes two or more types of distinct ADR procedures—e.g., mediation, arbitration, early neutral evaluation, summary jury or bench trials. If a district authorizes one distinct type of ADR, plus settlement conferences or ADR generally, it is counted in one of the "only" categories (e.g., ENE only). That said, some of the districts authorizing multiple types of distinct ADR may also include settlement conferences and/or a general authorization for ADR in their ADR rule; in fact, fifteen of the thirty-four districts do.

b. Where the word "only" is used in this table it means a district's written ADR documents mention only that one distinct form of ADR. Some of the districts that authorize only one of the distinct forms of ADR—e.g., mediation or early neutral evaluation—also include settlement conferences and/or a general authorization for ADR in their ADR documents.

c. "General authorization" means the district authorizes use of ADR or authorizes an "open track" or "general track" for ADR. These districts may mention specific forms of ADR, such as mediation and early neutral evaluation, but their written documents do not provide details that suggest authorization of a court-administered ADR program.

d. This arbitration-like process is authorized for certain types of cases arising under state law and uses a state tribunal.

e. Three districts authorize a combination of ADR types that could not be classified elsewhere.

Source: Stienstra, Federal Judicial Center Report 2011.

As evidenced above, mediation represents the most prevalent option among courts that implement only one form of ADR. Table 5.3 below, also evidences this trend, showing 67 per cent of the federal courts utilizing mediation in some form.

Notably, no District Court authorizes arbitration in isolation, but 23 districts offer arbitration as an option. The FJC Report notes that at the time of the study, 'only three of the ten mandatory arbitration districts continue to require use of arbitration for the full portion of their caseload that meets the statutory requirements; four others have made arbitration an ADR option, and three no longer authorize this procedure.'[48]

Table 5.3 Prevalence of ADR procedures

Type of ADR procedure authorized	Number and percent of district courts that have authorized the ADR procedure[a]	
	Number	Percent[b]
Mediation	63	67.0
Settlement conference	36	38.3
General authorization	27	28.7
Arbitration	23	24.5
Early neutral evaluation	23	24.5
Pro se mediation program[c]		
Non-prisoner pro se litigants	18	19.2
Prisoner pro se litigants	11	11.7
Summary jury or bench trial	14	14.9
Mini-trial	5	5.3
Case evaluation[d]	3	3.2
Settlement week[e]	2	2.1
Med/arb[f]	1	1.1

Notes:

a. A district is counted as authorizing a procedure if that procedure is mentioned as a distinct ADR procedure and not as one among several included in a general authorization to use ADR.

b. As a percentage of ninety-four district courts.

c. Many of the pro se mediation programs are new and experimental and not yet recorded in court local rules or other ADR documents. The information in this table comes from a survey of pro se services conducted by the Federal Judicial Center for the CACM Committee. Twenty-one districts have such programs, with eight districts offering both.

d. This arbitration-like process is authorized for certain types of cases arising under state law and uses a state tribunal.

e. During settlement week, the court's facilities are devoted to mediation of a roster of trial-ready cases. Attorneys from the district's bar serve as mediators.

f. In this procedure, a case first uses mediation and, if it does not settle, proceeds to arbitration.

Source: Stienstra, Federal Judicial Center Report 2011

Over 35 per cent of federal courts utilize (usually mandatory) settlement conferences. The local rules for the Northern District of California, for example, outline the procedure for settlement conferences:

> A judicial officer, usually a magistrate judge, helps the parties negotiate. Some settlement judges also use mediation techniques to improve communication among the parties, probe barriers to settlement and assist in formulating resolutions. Settlement judges might articulate views about the merits of the case or the relative strengths and weaknesses of the parties' legal positions. Often settlement judges meet with one side at a time, and some settlement judges rely primarily on meetings with counsel.[49]

Along with settlement conferences, ENE is a heavily used dispute resolution mechanism. Over a quarter of the districts make use of ENE in some fashion, although at the time of the FJC Report only three federal courts utilized ENE as the sole option available to litigants. Other, less prevalent forms of ADR authorized by the district courts include summary trials, mini-trials, case evaluation, settlement week, and med/arb.

Referral to ADR in the federal district courts

Generally, information about a federal court's referral process is in its ADR rules/ Local Rules and court website. The FJC Report groups federal court referrals into three categories:

- Mandatory referral of all cases to ADR;
- Referral to ADR with parties' consent; or
- Referral to ADR at the judge's discretion.[50]

The majority of courts that implement mediation allow the judge to refer the parties without consent, but are more divided when it comes to other ADR methods.[51] In the case of arbitration, federal districts tend to authorize voluntary use, rather than requiring party participation.[52] The FJC Report notes that 'districts may authorize different referral methods for different types of cases – for example, mandatory ENE for employment cases and voluntary ENE for other civil cases – and thus the number of districts authorizing the referral processes may be greater than the number of districts authorizing the ADR procedure.[53]

Table 5.4 Number of districts authorizing each referral process

ADR procedure (number of districts that authorize)	*Number of districts that authorize each type of referral process[a]*			
	Consent by all parties needed	*Judge May order without party consent*	*District mandates referral for all or specified cases*	*No information*
Mediation (63)	11	46	12	0
Arbitration (23)	11	9	4[b]	0
ENE (23)	7	13	5	1
General authorization (27)	13	13	0	2

Notes:

a. The total number of districts authorizing referral processes may be greater than the number of districts authorizing each type of ADR procedure because some districts authorize more than one type of referral process (for example, a district may authorize mandatory referral for some types of cases and voluntary referral for other types of cases).

b. Three of the original ten mandatory arbitration districts continue to require use of arbitration for all eligible cases; one includes it as an option in a program where use of some form of ADR is presumed.

Source: Federal Judicial Center Report 2011

Table 5.5 Types of neutrals available in federal court ADR programs

ADR procedure (number of districts that authorize)	Number of districts that authorize each type of neutral[a]				
	Judges	Court staff neutral	Panel of neutrals	Outside neutral	No information
Mediation (63)	4	9	42[b]	12	8
Arbitration (23)	-	-	21	2	1
ENE (23)	4	-	15	2	3
General authorization (27)	8	2	3[c]	13	7

Notes:

a. The total number of districts authorizing neutrals may be greater than the number of districts authorizing each type of ADR procedure because some districts authorize more than one type of neutral, depending on the type of ADR procedure authorized (for example, a district may authorize both a staff mediator and a panel of neutrals to provide mediation services).

b. Includes eight districts where the panel includes judges (in addition to attorneys and, in some districts, non-attorneys).

c. Includes one district where the panel includes judges (in addition to attorneys and, in some instances, non-attorneys).

Source: Federal Judicial Center Report 2011.

Neutrals in the federal court ADR system

Federal courts often offer information about authorized neutrals on their websites or in their local rules, including costs parties are expected to pay the neutrals, if any. Typically, in the context of mediation, arbitration, and ENE, a panel of neutrals is provided from which the parties choose. Each court has its own standards for qualification as a neutral, often requiring specific training. Some allow the use of outside neutrals, which gives parties more discretion.

Initially, court-connected ADR programs were provided without charge – many neutrals performed the services gratis in exchange for the exposure. As of 2011, however, parties are increasingly expected to pay for the neutrals' services, even in court-mandated ADR.[54] Some critics argue that forcing parties to cover costs in mandated ADR undermines the process, but there is little precedent rejecting payment requirements.[55] Some districts provide for reduced fees or fee waivers for qualifying litigants, and some districts require that neutrals offer pro bono services to be included on the district's panel of authorized neutrals.

Frequency of referrals to ADR in federal district court

The FJC was able to obtain referral information from only 49 federal district courts, which are not required to report the number of cases referred to or disposed of by ADR (summarized in Table 5.7). For their data, the FJC relied on the courts' submissions for ADR staffing supplements to the Administrative Office, as the applications provide a count of referrals.

Given the only data available is from district courts that may refer a greater number of cases to ADR than those that did not seek an increase in funding, these

Table 5.6 Compensation of neutrals in federal court ADR programs

ADR procedure (number of districts that authorize)	Number of districts that authorize each type of compensation arrangement[a]						
	Judges serve as neutrals	Court staff serve as neutrals	Party pays the fee	Non-court neutral serves pro bono	Court pays the fee	Tiered scheme[b]	No information
Mediation (63)	3	9	39[c]	6	-	11	5
Arbitration (23)	-	-	8	1	12	1	1
ENE (23)	3	-	11[d]	2	-	3	4
General authorization (27)	8	2	11	-	-	-	15

Notes:

a. The total number of districts authorizing a fee arrangement may be greater than the number of districts authorizing each type of ADR procedure because some districts authorize more than one type of fee arrangement, depending on the type of neutral used.

b. In a tiered scheme, the parties receive a small number of hours, typically between four and six, as a pro bono service. After this, the parties may continue the ADR process but must pay a fee to the neutral unless the neutral waives the fee. Some districts permit the neutrals to charge their market rates; others place limits on the fees that may be charged.

c. Includes seven districts that have set an upper limit on mediator fees.

d. Includes one district that has set an upper limit on neutral evaluator fees.

Source: Federal Judicial Center Report 2011.

numbers may be misleading. Further, the data does not account for the settlements that occur outside of the court-connected ADR program(s).

The data available indicates that mediation is the most frequently utilized form of ADR (17,833 referrals), while a significant number of cases are also referred to arbitration and ENE (2,799 and 1,320 referrals, respectively).

The FJC Report indicates that, as of 2011, 'the total number of cases referred to ADR in the applicant districts has ranged from the mid-to-high 20,000s,' and 'the ADR referrals represent 15 percent of the filed cases' in the 49 districts for which data was available.[57] It is unclear if these numbers are wholly accurate, given the fact that data was unavailable from all of the districts utilizing ADR programs, but it is clear that use of ADR in the federal courts continues to grow.

ADR in the Federal Circuit Courts of Appeals

The Federal Courts of Appeals also have their own court-connected ADR initiatives. A Federal Judicial Center publication on the use of mediation and conference programs in the federal courts of appeal recognized that 'mediation sessions exist to help parties communicate with one another, clarify their

Table 5.7 Referrals to ADR in federal district courts (reported)[56]
Number of cases referred to ADR in forty-nine federal district courts[a]
(Twelve-month period ending June 30, 2011)

ADR process	No. of cases
Mediation	17,833
Arbitration	2,799
CA-N multi-option program[b]	4,222
Early neutral evaluation	1,320
Settlement week	522
Summary jury trials	0
Mini-trials	0
Other[c]	1,571
Total	28,267

Notes:

a. Source: Applications to the Administrative Office of the U.S. Courts for supplemental funding for ADR staff.

b. In the submission from the Northern District of California, the number of referrals is reported as a total for the Multi-Option Program and is not broken down into the several different types of ADR offered under this program.

c. "Other" includes primarily judicially hosted settlement conferences.

Table 5.7 shows the number of ADR referrals for the twelve-month period ending June 30, 2011, which the FJC compiled using applications for 2012 funding.

Source: Federal Judicial Center Report 2011.

understanding of underlying interests and concerns, identify the strengths and weaknesses of legal positions, explore the consequences of not settling, and generate settlement options.'[58]

In 1974, only the Second Circuit Court of Appeals had a conference program. It would not be until 2005 that all of the appellate circuits had their own ADR programs. The following table, prepared by the Federal Judicial Center, demonstrates the year of implementation of Mediation and Conference Programs in the Federal Courts of Appeals.

In 1994, the Judicial Conference of the United States put into effect an amended Federal Rule of Appellate Procedure 33, which states:

> The court may direct the attorneys – and, when appropriate, the parties – to participate in one or more conferences to address any matter that may aid in disposing of the proceedings, including simplifying the issues and discussing settlement. A judge or other person designated by the court may preside over the conference, which may be conducted in person or by telephone. Before a settlement conference, the attorneys must consult with their clients and obtain as much authority as feasible to settle the case. The court may, as a result of the conference, enter an order controlling the course of the proceedings or implementing any settlement agreement.

Table 5.8 Year of implementation of mediation and conference programs in the federal courts of appeals

Second circuit	1974	Eleventh circuit	1992
Sixth circuit	1981	Fourth circuit	1994
Eighth circuit	1981	Seventh circuit	1994
Ninth circuit	1984	Third circuit	1995
D.C. circuit	1987	Fifth circuit	1996
Tenth circuit	1991	Federal circuit	2005a
First circuit	1992		

Notes:

a. From 1989 to October 3, 2005, the Federal Circuit had a settlement program. The court replaced this with a mediation program in 2005. Under the settlement program, the court's local rule required that attorneys for the parties hold settlement discussions in all civil cases that met eligibility requirements. Generally, the court did not keep track of attorneys' compliance with this rule or learn with certainty whether the settlement discussions that attorneys did hold were substantive or merely pro forma.

Source: Federal Judicial Center Report 2011.

As court-ordered mediation does not serve to toll the deadlines for submission of briefs, etc., appellate mediation is generally scheduled shortly after the appeal. This is due in part to the belief that litigants' willingness to settle is decreased as the appeal progresses due to the expense incurred. Moreover, even where unsuccessful, mediation may benefit counsel in the briefing process by clarifying the core issues.[59]

Although many federal appellate courts have similar programs, there are some jurisdictional differences that were the subject of a Federal Judicial Center report. As a preliminary matter, the names utilized for the programs vary. Initially, most federal appellate courts used the title 'preargument conference program,' but now utilize names such as 'mediation program' or 'settlement/conference program.'[60] In all but the Eighth and Federal Circuits, the settlement conferences are mandatory, although some allow the removal of a case from the program at a party's request or at the neutral's discretion.[61]

Federal appellate courts also differ in the type of cases they refer to ADR. Some refer almost all civil appeals to ADR, while others confine ADR to specific categories of cases. For example, in the First and Second Circuits, nearly all civil cases are scheduled for a civil appeals management conference, and in the Eleventh Circuit, all fully counseled civil appeals (except prisoner, habeas corpus, and immigration appeals) are eligible for mediation. The other circuits, however, schedule mediation 'only in cases that appear likely to achieve settlement on some or all of the issues on appeal.'[62] The Federal Judicial Center noted that appellate mediation offices generally do not schedule pro se cases for mediation out of concern that pro se parties might mistake the mediator as an advisor or feel forced to settle.[63]

Per the Federal Judicial Center's report, most programs hold telephone conferences. In the Second, Fourth, Fifth, Sixth, Seventh, Eighth, Ninth, and Tenth Circuits, neutrals indicated that they conduct from 50 per cent to 95 per cent of their sessions by telephone. This aids in reducing travel expenses – as most federal appellate courts are not located in the same forum as the litigants. In the

First, Third, and D.C. Circuits, most mediation sessions are held in person, but travel expenses or other factors may require telephone conferences.[64] The initial mediation sessions in the Federal Circuit are always held in person.

In nearly all of the programs, the Federal Judicial Center found that attorneys employed by the court conduct the mediations.[65] In some courts, senior federal judges or retired state judges conduct the conferences, while the District of Columbia and Federal Circuits are the only circuits that use volunteer mediators.[66]

ADR in the US state courts

Like US federal courts, US state courts receive a startling number of claims per year. The National Center for State Courts (NCSC) periodically reports on state court caseloads and indicated 103.5 million of incoming cases in 2010, the majority of which were in courts of limited jurisdiction (66 per cent), such as traffic courts, while the remaining cases were in general jurisdiction (18 per cent) and single-tiered courts (17 per cent).[67]

The courts in the US are overburdened and it can often take years for a complex case to get to trial. Understandably, the percentage of civil cases that actually proceed through trial is staggeringly low.[68]

The NCSC obtained data from reporting courts for the years 2012–2013 regarding the number and percentage of civil cases that proceeded to jury trial, which reveals that – on average – less than *half of a percent* of state court civil cases proceed through to a jury trial. While it is unclear from the statistics provided if case disposition was achieved by ADR, bench trial, or another method, the figures demonstrate the increasing proclivity among US litigants for avoiding trial.

One possible explanation for the number of cases that never reached trial is the prevalence of ADR in the modern state courts. As of writing, only nine US states and the District of Columbia lacked dedicated dispute resolution statutes or rules.

Table 5.9 Total incoming caseloads reported by State Courts 2010 by case category (in millions)

Case category	Single-tiered		General	Single-tiered + general		Limited		Total	Percent of total
Traffic	11.1	+	3.3	= 14.4	+	41.9	=	56.3	54%
Criminal	2.6	+	3.7	= 6.3	+	14.1	=	20.4	20%
Civil	2.5	+	6.6	= 9.2	+	9.8	=	19.0	18%
Domestic relations	.7	+	3.5	= 4.2	+	1.7	=	5.9	6%
Juvenile	.2	+	.9	= 1.2	+	.7	=	1.9	2%
Total incoming	17.2	+	18.1	= 35.3	+	68.2	=	103.5	
Percentage of total	17%	+	18%	= 34%	+	66%	=	100%	

Source: National Center for State Courts.

Table 5.10 Total incoming caseloads reported by State Courts 2010 (in millions)

State	Tier[a]	2012 and 2013 total civil dispositions		2012 and 2013 civil jury trials		2012 and 2013 civil jury trial rate	
Alabama	General	43,691	N/A	347	N/A	.79%	N/A
Alaska	General	N/A	7,469	N/A	23	N/A	.31%
California	Single	1,006,956	921,830	1,737	1,668	.17%	.18%
Connecticut	General	133,302	135,2446	286	295	.21%	.22%
Florida	General	421,460	465,495	954	939	.23%	.20%
Hawaii	General	N/A	9,309	N/A	12	N/A	.13%
Indiana	General	414,424	420,899	291	248	.07%	.06%
Iowa	Single	135,451	N/A	204	N/A	.15%	N/A
Kansas	General	160,711	149,933	86	85	.05%	.06%
Kentucky	General	61,234	51,633	148	142	.24%	.28%
Michigan	General	64,590	61,650	260	288	.40%	.47%
Minnesota	Single	182,470	168,157	320	306	.18%	.18%
Missouri	General	292,412	276,001	405	398	.14%	.14%
Nebraska	General	N/A	11,223	N/A	70	N/A	.62%
New Jersey	General	849,826	923,239	1,428	1,396	.17%	.15%
Nevada	General	52,869	45,183	213	162	.40%	.36%
New York	General	177,457	N/A	2,048	N/A	1.15%	N/A
Ohio	General	169,329	148,523	568	560	.34%	.38%
Pennsylvania	General	147,163	151,968	1,036	838	.70%	.55%
Texas	General	213,389	210,367	1,116	987	.52%	.47%
Utah	General	101,348	96,609	79	99	.08%	.10%
Washington	General	148,707	140,622	285	310	.19%	.22%
West Virginia	General	30,559	30,567	106	99	.35%	.32%

Notes:

a. The term "trial court" encompasses "single-tiered courts, courts of general and limited jurisdiction, and courts of special jurisdiction (e.g., water court, probate court, and small claims court), regardless of whether they hold trials or not." Courts of general jurisdiction encompass many courts—including superior court, circuit court, district court, court of chancery, court of common pleas, etc.—but are "the highest trial court in the state for the matters they hear." State Court Guide to Statistical Reporting, National Center for State Courts, Court Statistics Project, Version 2.1.1, 2 (October 30, 2015).

Source: National Center for State Courts.

Although the federal court system in the US is vast, it cannot compare with the size of the US state court system. Every state has its own system of lower courts, appellate courts, and a Supreme Court. While some states have a limited number of courts, others have almost as many courts as there are US states (sometimes more).[69] Given this fact, an analysis of each state's ADR programs is impractical for the purposes of this chapter. Instead, we will focus on the State of Maryland, which has a well-established ADR program and which has conducted research on not only the efficacy of its own court-connected ADR programs, but also court-connected ADR programs in other states.

Within the Maryland court system district courts, circuit courts, and the Court of Special Appeals each have their own independent ADR programs.[70] Maryland Judiciary's Administrative Office of the Courts also has a Mediation and Conflict Resolution Office (MACRO), which 'serves as a dispute resolution resource for the state, supports and offers technical assistance to conflict resolution programs in the courts and community, and promotes appropriate dispute resolution in every field.'[71] MACRO works to help establish, promote quality assurance, and evaluate ADR services and education throughout Maryland.[72]

Reports on ADR programs in Maryland district and circuit courts

The Statewide Evaluation of ADR in Maryland ('*MD Evaluation*'),[73] studies and polls participants in Maryland's court-connected ADR programs in order to 'assess the costs, benefits, and effectiveness of ADR options offered with the Maryland court system.'[74] This initiative is led by a team composed of members of the Center for Conflict Resolution at Salisbury University,[75] the Center for Dispute Resolution at the University of Maryland Francis King Carey School of Law (C-DRUM),[76] Community Mediation Maryland,[77] the Institute for Governmental Service and Research at the University of Maryland,[78] and the Maryland Administrative Office of the Courts.[79]

Another resource available to practitioners and litigants through the Statewide Evaluation of ADR in Maryland is a report entitled *Alternative Dispute Resolution Landscape: An Overview of all ADR in the Maryland Court System* (the '*ADR Landscape*'). The information in the *ADR Landscape* was gathered and compiled between July 2010 and January 2013, from interviews with ADR Coordinators and courthouse staff throughout Maryland, and provides a 'comprehensive overview of ADR in Maryland.'[80]

An additional resource offered by the Statewide Evaluation of ADR in Maryland is a report on ADR in the Maryland district courts entitled, *Impact of Alternative Dispute Resolution on Responsibility, Empowerment, Resolution, and Satisfaction with the Judiciary: Comparison of Short- and Long-Term Outcomes in District Court Civil Cases*.[81] In connection with the evaluation, surveys were used to draw a 'comparison between individuals who used ADR (the 'treatment' cases) and those who went through the standard court process without ADR (the 'control' cases).'[82]

ADR in Maryland district courts

The *MD Evaluation* provides the following summary of the civil ADR program in Maryland's circuit courts:

> The District Court ADR Office started in Anne Arundel County in 1998 and currently operates in 63 per cent (15 out of 24) of jurisdictions in Maryland. The ADR programs are managed by the District Court ADR Office whose

mission is to 'establish and maintain high quality ADR programs that empower litigants.' Effective January 1, 2013, Chapter 300 of Title 17 of the Maryland Rules addresses ADR proceedings in the District Court. As such, the ADR Office sets statewide practices for the provision of quality ADR services pursuant to the Maryland Rules.

The District Court ADR Office is administered by an executive director and deputy director and seven additional staff members. The program operates with funding from the Maryland Judiciary and volunteer services from ADR practitioners and partnerships with community mediation centers and the University of Maryland Carey School of Law Mediation Clinic. ADR practitioners within the District Court ADR program provide mediation or settlement conferencing services. In all instances ADR is provided at no charge to the litigants.[83]

Almost every Maryland district offers 'day of trial' mediation and settlement conference programs, and some also offer civil pretrial mediation and settlement conference programs.[84] In civil pretrial mediation, litigants are referred to partner community mediation centers. In civil pretrial settlement conferences, the parties meet with a court-provided settlement conference neutral. The Consumer Guide to ADR Services in Maryland identifies the following differences between district and circuit court settlement conferences:

- The parties' attorneys do most of the negotiating during a settlement conference and, unless they lack an attorney, parties are generally less involved in the settlement conference process than in mediation.
- In a settlement conference, parties are more likely to remain in separate rooms, with the neutral caucusing with and transmitting communications from the various parties.
- Settlement conference facilitators are more likely to evaluate and make recommendations for how to settle the case.[85]

If resolution does not occur after the mediation or settlement conference, the case proceeds to trial. On the day of trial, the parties have the option to mediate or participate in a settlement conference. In civil day of trial mediation, the parties may request a mediator or a settlement conference from the court and, if the parties are still unable to settle the dispute, the trial will proceed.

Circuit courts throughout Maryland provide mediation and settlement conferencing services for most civil cases.[86] Generally, the costs of such services are shared by the parties. Although the court may refer the parties to a roster of approved mediators, the parties may – with consent of all parties to the suit – elect to use a different mediator of their own choosing.[87]

Maryland appellate courts also have an ADR initiative. Maryland Appellate Rule 8-206 provides the court with procedures for 'ADR and Scheduling Conferences.' The court will determine whether the parties and their attorneys must participate in a pre-hearing conference or mediation session. A judge who conducts a pre-hearing conference will not sit on the panel of judges assigned to hear the appeal.

In January 2010, the Maryland Court of Special Appeals (COSA) created a pilot mediation conference program and the COSA ADR Division. The COSA ADR Division, which has a full time staff, offers mediation conferences under MD Rule 8-206(b) and is responsible for screening cases for ADR, scheduling the ADR session, conducting the ADR process, and quality control.

Per the *ADR Landscape*, as of 2013, approximately 10–12 per cent of the civil appeals filed with the Maryland Court of Special Appeals each year go through a pre-hearing conference, by order of the court or request of the parties, and approximately 110 appeals are resolved through mediation or settlement conference.[88] If a party fails to attend the conference, the court may issue sanctions against that party, which may include: (1) dismissal of part or all of the appeal, (2) assessment of the reasonable expenses caused by the failure, including attorneys' fees, against the party or attorney, (3) assessment of part or all of the appellate costs against the party or attorney, or (4) imposition of any other 'appropriate sanction.'[89]

The Maryland courts' empirical inquiry into the costs and benefits of court-connected ADR programs is not limited to Maryland's court systems. In connection with its relatively new ADR program, the Maryland Court of Special Appeals ADR Division (COSA ADR), in conjunction with the Center for Dispute Resolution at the University of Maryland (C-DRUM), conducted a survey of appellate mediation programs nationwide and wrote a report based on their findings entitled *An Analysis of the Maryland Court of Special Appeals ADR Division January 2012 Appellate Mediation Program National Questionnaire.*[90]

The analysis examined each responding jurisdiction's program, number of years in operation (through 2011), method of authority, mandatory/ordered mediation, mediation for pro se individuals, use of co-mediation, the number of civil appeals, number of cases mediated, number of case settled (full), and number of cases settled (partial).

The data collected indicated that the 35 programs with more than 20 mediations per year produced an average resolution rate in court-connected ADR of 54.33 per cent.[91] In the 28 programs with more than 50 mediations per year, the average resolution rate in court-connected ADR of 45.46 per cent.[92] Of the 40 state appellate programs analyzed, only 15 programs indicated that they conduct over 100 mediations and have a mandatory or voluntary/upon request (or both) appellate mediation programs similar to Maryland. Like Maryland, eight other appellate mediation programs were initiated within the five years preceding the report.[93]

The future of US court-connected ADR programs

Twenty years have passed since Congress mandated development of court-connected ADR programs and ADR is now an established part of many courts' regular case management practices.

Criticism of court-connected ADR

Lack of free will

The mandatory nature of court-connected programs has attracted criticism because it seemingly 'forces' litigants into a non-traditional process that they have not chosen. The perception of 'compulsion and coercion' also raised criticism and concern early on in the US courts' experiments with connected ADR.[94] Critics painted a nefarious picture – often implying that voluntary ADR programs were 'imposed through persuasive arguments of judges or others who promise escape from the real or perceived disadvantages of traditional litigation.'[95]

Critics note that parties lack the ability to appeal a court's order compelling mediation and the negative consequences may result from non-compliance. In *Short Bros. Constr., Inc. v. Korte & Luthohan Contractors, Inc.*,[96] the court determined that an order to mediate was an act of docket management and was not an injunctive order. Consequently, the order to mediate was not appealable. Moreover, courts sanction parties that do not participate in court-connected mediation in 'good faith.'[97]

There have been exceptions, however. In the 1987 case *Strandell v. Jackson*, the Seventh Circuit Court of Appeals held that a court does not have the authority to require parties to participate in a summary judgement trial, a form of ADR.[98]

Subsequently, the group that spearheaded the revision of Federal Rule of Civil Procedure 16 rejected the *Strandell* court's interpretation. Under Rule 16(c)(9) of the Federal Rules of Civil Procedure, the courts are authorized to consider and take 'appropriate action with respect to 'settlement and the use of special procedures to assist in resolving the dispute when authorized by statute or local rule.' The last sentence of Rule 16(c) provides that '[i]f appropriate, the court may require that a party or its representative be present or reasonably available by telephone in order to consider possible settlement of the dispute.'[99]

In both federal and state court, parties may have to face sanctions for bad faith. Such sanctions may include imposing court costs, awarding attorneys' fees, contempt, denial of trial de novo (amounting to confirmation of an arbitrator's award), and dismissal of the pending litigation.[100] Such sanctions have been imposed for a variety of reasons, including, but not limited to, refusal to attend or participate in ADR sessions, and non-compliance with decisions or settlements reached in ADR proceedings.[101]

A limited number of courts have held, however, that the confidential nature of mediation precludes parties from pursuing these bad faith actions in circumstances where doing so would require disclosure of a party's conduct in mediation. In *Foxgate Homeowners' Ass'n, Inc. v. Bramalea California, Inc.*, the Supreme Court of California reversed the trial court's order granting sanctions for bad faith mediation on the grounds that none of the confidentiality statutes made an exception for reporting bad faith conduct or for imposition of sanctions when doing so would require disclosure of communications or a mediator's assessment of a party's conduct.[102]

A party's refusal to abide by an agreement or order obtained in ADR proceedings may face sanctions, but this is not a strict rule. Courts have imposed sanctions on an attorney for continuing to prosecute a suit after pronouncement of an arbitral award,[103] but sanctions have been denied where the plaintiff, after settling a suit, amended his complaint to allege a breach of a mediation agreement.[104] 'The types of sanctions approved for use in situations involving enforcement of agreements reached in a dispute resolution proceeding include discovery sanctions, contempt citations, striking the pleadings, and other specific sanctions provided by trade association rules.'[105]

Each jurisdiction has its own established case precedent regarding ADR and a wise litigant will familiarize himself/herself with the approach taken by a given jurisdiction in order to avoid violating any local precepts governing bad faith conduct in ADR proceedings.

This issue can be further complicated where the parties to a construction project already have an ADR clause in their agreement. As one commentator noted:

> If the litigation is of a type and in a court in which ADR is mandatory, the judge may require participation in a court-connected ADR program even where the parties have already been through a private, non-binding ADR process governed by the contract. A contractual ADR clause may express the parties' intention that the contractual process replace a court-ordered process. However, there is no guarantee that a court will honor that intention.[106]

The author noted that a binding arbitration clause, which takes the parties out of the court system, is a way to avoid this issue, but for parties that want their day in court, that is not a satisfactory 'solution.'

Misnomer of classification

Another of the primary criticisms levied against court-connected mediation is that in its current application it resembles traditional pretrial settlement conferences, not traditional mediation. As one authority noted,

> While what is being called 'mediation' in the courts may encompass the interest-based, problem-solving, or relational approaches, which mediation advocates envisioned 15 or 20 years ago, the combination of increased participation by lawyers and the close connection with litigation of court-referred mediation cases is leading to the increased 'legalization' of mediation. Court-referred clients often believe the desired outcome that propelled them to court initially will be met. Their misplaced assumptions about the type of process being ordered and the degree of court oversight can lead to disappointment with the process, the outcome, and the courts in general. Similar disappointment can result if promises that mediation is 'faster, cheaper, and better' are not met. Perhaps most dangerous, the blurring of boundaries in the court's roles can lead to confusion and leave room for the possibility of coercion.[107]

Some express the concern that court-connected mediation 'capitulat[es] to the courts' routine' and has 'lost sight of the core values of mediation and simply become absorbed into the courts' traditional methods of adversarial dispute resolution.'[108] An article examining the problems with modern court-connected dispute resolution, *ADR in the Courts: Progress, Problems and Possibilities*, articulated the following 'signs of capitulation':

- Limited allotments of time for mediation;
- Attorneys or other representatives attending without the litigant;
- Attorneys excluding clients from the process;
- Attorneys advising clients not to participate;
- Mediators suggesting caucuses between attorneys without their clients; and
- Failure to assign trained mediators as neutrals.[109]

The authors of this article suggest a return to the 'core values' of mediation, such as a means to provide efficient resolution of conflict, to preserve relationships between parties to a dispute, and provide each side an opportunity to be heard. They proffered strategies for improving court-connected ADR to avoid this result, including: improved education on court-run ADR process for litigants, attorneys, and neutrals; clear delineation between court and mediation procedures; and implementation of quality control programs.[110]

Moreover, there have been genuine concerns that compulsion of mediation can lower the admirable settlement rate of voluntary mediation. A failed mandatory mediation will even add time and cost consumed to resolve the disputes, so without a higher success rate, it is unattractive to make mediation mandatory.

Benefits of court-connected ADR in the construction context

Efficiency, reduced cost, and confidentiality have always been private ADR's most appealing advantages. While court-connected ADR in many jurisdictions is still in its infancy, proponents would argue that the introduction of court-connected ADR added 'judicial economy' to that list of benefits.

Although case backlogs may be the primary impetus for court-connected ADR, the American notion of self-determination has contributed to its rapid growth and success. The programs provide 'a neutral platform for parties to exchange information and evaluate their risks and their expected future time and expenses in a given case, with an eye toward reaching a reasonable compromise when it is in their best interests.'[111] Moreover, one commentator notes that mediation in litigation and arbitration can be 'an effective case management process by which to reach consensus of the parties on deposition discovery, scheduling, hearing time allotted to each party, and how the case otherwise might most efficiently be tried.'[112]

In her article on the use of mediation in the state appellate system, one expert offers the following justifications for engaging in the process:

- Conservation of judicial resources affordability procedurally fair process promotes compliance global settlements are possible obviates the risk of appeals, and
- Earlier resolution.[113]

Construction counsel are learning to appreciate the benefits of court-connected ADR. Parties that may have been reluctant to submit to mediation are now compelled to participate in good faith, permitting settlement of cases that would otherwise have remained embroiled in litigation. Even in situations where parties do not settle their disputes, the presence of a neutral – a former judge, for example – can assist parties in narrowing the outstanding issues for trial. Moreover, where neutrals are provided by the court, such as in Maryland state courts, parties that otherwise could not afford to mediate may enjoy its benefits.

Although some still raise concerns about the perceived failings and risks of court-connected ADR, their comments lack the bombast of opponents in the 1970s-1990s. Some of this is no doubt due to the fact that many courts, via programs such as MACRO, have implemented procedures that seek to address the issues at the root of these concerns. As courts (and counsel) become savvier and the ADR programs themselves more tailored, it is likely the detractors will continue to decrease.

Conclusion

The construction industry s is in many ways responsible for the development of ADR in the US The interconnected nature of the industry spurred the need to prioritize conciliatory efforts over traditional litigation. The two primary forms of ADR, arbitration and mediation, are still the most common in the US, but a number of other methods exist. Distinct federal and state court systems have developed a panoply of administered ADR options, many of which are mandatory, and experienced construction practitioners are expected to maintain familiarity with the programs of their local courts.

Overburdened by litigation, most state and federal courts maintain ADR programs. Many of the federal courts have established ADR programs relatively recently, following the Congressional passage of the Alternative Dispute Resolution Act of 1998. Their growth has been significant, and 67 per cent of the federal courts – and 100 per cent of the Federal Courts of Appeals – now utilize ADR in some form or another. A substantial number of these federal ADR programs are mandatory.

Practices of this sort and their obligatory nature have been the focus of some criticism, with critics particularly criticizing the removal of a party's ability to opt out of the process. Critics cite the cost to the parties, who are often required to compensate not only their counsel for time invested in complying with court-mandated ADR protocols, but also neutrals. The utility of ADR processes, which sometimes resemble traditional pretrial settlement conferences, in situations in which non-binding negotiation has already failed has also been questioned.

As US courts and practitioners grow more familiar with the process, however, construction participants appreciate the relative benefits of US court-connected

ADR, such as increased efficiency, reduced overall cost, and confidentiality. Moreover, as courts continuously refine the ADR programs and address deficiencies, criticisms are bound to decline.

Acknowledgements

Special thanks are given to Donna Stienstra of the Federal Judicial Center, Mala Malhotra-Ortiz, Esquire and Scottie Reid, Esquire, of the Maryland Court of Special Appeals' Alternative Dispute Resolution Division, and Alexander Gorelik, a student at George Washington University School of Law, for their contributions during the preparation of this chapter.

Notes

1 Richard C. Reuben, *Constitutional Gravity: A Unitary Theory of Alternative Dispute Resolution and Public Civil Justice*, 47 UCLA L. Rev. 949, 972 (2000).

2 See *Wimsatt v. Superior Court*, 152 Cal. App. 4th 137, 61 Cal. Rptr. 3d 200 (2d Dist. 2007) (holding that mediation briefs and e-mails quoting same were privileged); *Paranzino v. Barnett Bank of South Florida, N.A.*, 690 So. 2d 725 (Fla. 4th DCA 1997), cause dismissed (Fla. June 2, 1997) (dismissing an action after plaintiff breached confidentiality by revealing details of mediation to the media).

3 See 7 Bruner & O'Connor on Construction Law, *General Principles of Mediation*, § 21:312 (May 2015) (citing UMA, Prefatory Note (2001)).

4 National Conference of Commissioners on Uniform State Laws (NCCUSL), Uniform Mediation Act §§ 1 to 16 (2001).

5 Uniform Mediation Act § 3.

6 See www.uniformlaws.org.

7 See 7 Bruner & O'Connor on Construction Law, *Genesis of construction arbitration*, § 21:1 (May 2015).

8 9 USC. § 2.

9 See 7 Bruner & O'Connor on Construction Law, *Historical development of arbitration – From pariah to prince*, § 21:2 (May 2015).

10 9 USC. § 1, *et seq*. The FAA does not always govern, however. For example, the FAA is inapplicable in cases involving the Taft-Hartley Act, Section 301 of the Labor-Management Relations Act. *Brown v. Witco Corp.*, 340 F.3d 209, 220 (5th Cir. 2003).

11 NCCUSL, Revised Uniform Arbitration Act (2000), available at www.uniformlaws. org/Act.aspx?title=Arbitration%20Act%20(2000).

12 As of this writing, states that have enacted the RUAA include Alaska, Arizona, Arkansas, Colorado, District of Columbia, Florida, Hawaii, Michigan, Minnesota, Nevada, New Jersey, New Mexico, North Carolina, North Dakota, Oklahoma, Oregon, Utah, Washington, and West Virginia. See www.uniformlaws.org; see also Legal Information Institute, *Alternative Dispute Resolution – State Laws*, available at www.law.cornell.edu/wex/table_alternative_dispute_resolution.

13 www.adr.org.

14 www.cpradr.org.

15 www.jamsadr.com.

16 See www.adr.org.

17 In disputes totaling over $100,000, AAA arbitration participants must mediate their claims concurrent with the arbitral process, although the parties may waive the requirement.

18 See 2007 Revisions to AIA Contract Documents, available at www.aia.org/groups/aia/documents/document/aiab078763.pdf.

19 Most recently, SCOTUS ruled that, under the FAA, arbitration agreements in company contracts must be honored, regardless of whether there are more consumer-friendly protections set by states. *DIRECTV, Inc. v. Imburgia, et al.*, 135 S. Ct. 463 (2015).

20 Abramowitz, Ada J., *The Keys to Keeping a Project on Track*, 8 No. 2 Journal of the American College of Construction Lawyers 3 (August 2014).

21 Stipanowich, Thomas J., *The Multi-Door Contract and Other Possibilities*, 13 Ohio St. J. on Disp. Resol. 303, 378 (1998).

22 Abramowitz, *supra* note 24.

23 Hafer, Randy, et al., *Dispute Review Boards and Other Standing Neutrals: Achieving 'Real Time' Resolution and Prevention of Disputes*, CPR Dispute Prevention Briefing: Construction, 8 (2010).

24 Hafer, *supra* note 28 at 5; see also Construction Industry Institute, IR23-2 – Prevention and Resolution of Disputes Using Dispute Review Boards, available at https://www.construction-institute.org/scriptcontent/more/ir23_2_more.cfm.

25 See Article 12 of ConsensusDOCS 200-2007; AIA form A201-2007, §§ 1.1.8, 14.2.2 and 15.2; see also Mark H. McCallum, *Getting Directly to the Point of the Contested Matter: Dispute Mitigation & Resolution in ConsensusDOCS Construction Forms*, American Bar Association Forum on the Construction Industry, 17 (September 11–12, 2008)

26 7 Bruner & O'Connor Construction Law § 21:3 (referencing James Groton, The Progressive or 'Stepped' Approach to ADR: Designing Systems to Prevent, Control, and Resolve Disputes, in Construction Dispute Resolution Handbook (1997)); see also 2007 ConsensusDOCs 200 General Conditions, Article 12; Peter C. Halls, *General Conditions: A Comparison of ConsensusDOCS and the Revised AIA Documents*, Minnesota CLE (May 21, 2008).

27 7 Bruner & O'Connor Construction Law § 21:3 (citing Nael G. Bunni, *The FIDIC Forms of Contract 460* (3d ed. 2005)).

28 *Id.*

29 See 42 USC. §7401, et seq. (1970); visit www.EPA.gov.

30 *Black's Law Dictionary* (10th Edition 2014).

31 28 USC. § 1491(a)(1).

32 Taken from www.supremecourt.gov/faq.aspx

33 Court-referred programs fall between court-connected and private ADR, and allow independent mediators to assist the process. The private mediator must meet the standards for court-appointed mediators, and, despite being independent, often receive funding and referrals from the court.

34 National Standards for Court-Connected Mediation Programs (1999), available at: www.caadrs.org/downloads/NationalStandards.pdf.

35 National Standards for Court-Connected Mediation Programs 4.1.

36 *Id.* at 6.1, 6.5, 6.6, and 16.1.

37 *Id.* at 5.2.

38 *Id.* at 5.1.

39 *Id.* at 2.0 and 8.1.

40 Judicial Improvements and Access to Justice Act, Pub. L. No. 100-702, 102 Stat. 4642 (1988) (codified at scattered sections of 28 USC.A. §§ 651 to 658) (as amended by section 1 of Public Law 105-53); see also Rich, Alan B., Mathvin, Gaynell C., and Crisman, Thomas L., *The Judicial Improvements and Access to Justice Act: New Patent Venue, Mandatory Arbitration, and More*, Berkeley Technology Journal, volume 5, issue 2 (September 1990).

41 28 USC. § 471; but see 28 USC.A. § 652(b) (prohibiting orders to arbitrate in any action based on alleged violations of constitutional rights or federal civil rights statutes).

42 See 28 USC. § 471(a)(6) and § 476.

43 Stienstra, Donna, *ADR in the Federal District Courts: An Initial Report*, 1–2 (November 16, 2011). The CACM committee also prepared guidelines to assist the courts in developing ADR programs, which were not adopted as policy or officially distributed. The CACM committee also prepared guidance regarding fees paid to ADR neutrals, which the Judicial Conference issued as policy. *Id.*

44 The legislation that prompted the Civil Justice Reform Act expired in 1997. See *In re Atl. Pipe Corp.*, 304 F.3d 135, 141, n. 2 (1st Cir. 2002) (citing Tobias, Carl, *Did the Civil Justice Reform Act of 1990 Actually Expire?*, 31 U. Mich. J. L. Reform 887, 892 (1998).

45 Stienstra, *supra* note 60, at 2.

46 The Federal Judicial Center prepared the report based on a 'review of local rules, general orders, CJRA plans, internal operating procedures, web sites, and any other written source [found] that describes a district's ADR procedures.' Stienstra, *supra* note 60, at 4.

47 Stienstra, *supra* note 60, at 5.

48 Stienstra, Donna, *supra* note 60, at 6.

49 See Local Rule 7, United States District Court for the Northern District of California.

50 Stienstra, *supra* note 60, at 8.

51 Federal case law has upheld a judge's unilateral right to order parties to mediation. See *In re Atl. Pipe Corp.*, 304 F.3d 135 (1st Cir. 2002)(district court has inherent power to order non-consensual mediation).

52 Stienstra, *supra* note 60, at 9.

53 Stienstra, *supra* note 60, at 8.

54 Stienstra, *supra* note 60, at 11.

55 The case, *Jeld-Wen, Inc. v. Superior Court*, 146 Cal.App.4th 536, 53 Cal.Rptr.3d 115 (2007), offers a rare example of a court overturning an order directing a party pay neutral fees. In *Jeld-Wen*, the state court issued an order mandating parties to attend and pay for private mediation sessions. When an uninsured party failed to attend, that party was ordered to attend the next session, and a monetary sanction was imposed. The appellate court held that the trial court lacked authority to compel attendance and payment for private mediation.

56 Table 7 shows the number of ADR referrals for the twelve-month period ending June 30, 2011, which the FJC compiled using applications for 2012 funding.

57 Stienstra, *supra* note 60, at 15.

58 Niemic, Robert J., *Mediation and Conference Programs in the Federal Courts of Appeals: a sourcebook for judges and lawyers* (Second Edition, Federal Judicial Center 2006)

59 Niemic, *supra* note 78, at 10.

60 *Id.*, at 7.

61 *Id.*, at 11.

62 *Id.*, at 9.

63 *Id.*, at 8. Attorneys attending the settlement conferences without their client must have settlement authority. *Id.* at 12.

64 The Federal Judicial Center notes that these are 'relatively compact' circuits, making in-person conferences less difficult to plan and attend. Niemic, *supra* note 78, at 10–11.

65 These attorneys have a variety of titles, including circuit mediator, settlement counsel, conference attorney, and staff counsel. Courts report that most had prior experience or training in mediation and negotiation techniques. Niemic, *supra* note 78, at 13.

66 *Id.* at 13.

67 LaFountain, R., Schauffler, R., Strickland, S. & Holt, K., Examining the Work of State Courts: An Analysis of 2010 State Court Caseloads, 3 (National Center for State Courts 2012).

68 The State Court Guide to Statistical Reporting defines a Jury Trial as being counted when 'the jury has been sworn, regardless of whether a verdict is reached.' See www. ncsc.org.

69 Alabama, for example, has 41 judicial districts (the general jurisdiction trial court), civil and criminal court of appeals, and a supreme court. Kentucky has 57 circuit courts, a court of appeals, and a supreme court. California has 58 county courts, six appellate districts, and a supreme court. See Ron Malega, PhD, and Thomas H. Cohen, J.D., PhD, *State Court Organization, 2011*, Special Report of Bureau of Justice Statistics (November 2013).

70 For general information on the Maryland Courts' ADR programs, see www.courts. state.md.us/legalhelp/mediationadr.html.

71 See *Consumers' Guide to ADR Services in Maryland*, Maryland Mediation and Conflict Resolution Office, 9 (June 2015).

72 See *Id.*

73 For more information visit their website at www.marylandadrresearch.org.

74 Statewide Evaluation of ADR in Maryland: Research on Alternative Dispute Resolution in the Maryland Judiciary, available at www.marylandadrresearch.org.

75 See www.conflict-resolution.org.

76 See www.law.umaryland.edu/programs/cdrum/.

77 See www.mdmediation.org/.

78 See www.igsr.umd.edu/.

79 See www.courts.state.md.us/aoc/.

80 Report is available at: www.marylandadrresearch.org/landscape.

81 Charkoudian, Lorig, et al. *Impact of Alternative Dispute Resolution on Responsibility, Empowerment, Resolution, and Satisfaction with the Judiciary: Comparison of Short- and Long-Term Outcomes in District Court Civil Cases*, Community Mediation Maryland, State Justice Institute (February 2016).

82 *Id.* at 1.

83 The District Court ADR Program website contains educational materials, a roster of authorized neutrals, and its newsletter. See *Alternative Dispute Resolution Landscape: An Overview of all ADR in the Maryland Court System*, Part One: A Statewide Perspective (Spring 2014), available at www.marylandadrresearch.org/ landscape/part1/district-court-of-maryland.

84 See *Consumers' Guide to ADR Services in Maryland*, Maryland Mediation and Conflict Resolution Office (June 2015); see also www.courts.state.md.us/district/adr/ what.pdf.

85 See *Consumers' Guide to ADR Services in Maryland*, Maryland Mediation and Conflict Resolution Office, 13 (June 2015).

86 In the absence of a court referral, court-administered mediation is also available upon request and parties may always pursue private mediation. See *Alternative Dispute Resolution Landscape: An Overview of all ADR in the Maryland Court System* (Spring 2014)(internal citations omitted), available at www.marylandadrresearch. org/landscape/part1/circuit-courts-of-maryland.

87 See *Consumers' Guide to ADR Services in Maryland*, Maryland Mediation and Conflict Resolution Office, 12 (June 2015) Courts often prohibit referral of certain cases to the district and civil courts' ADR programs. Maryland Rule 9-205(b), for example, prohibits referral of child custody disputes where abuse has been alleged, and Rules 17-201(b) and 17-302(b) prohibit referral of protective order actions. Many states have similar laws in effect. See, *e.g.*, Ala. Code § 6-6-20 (2007); Alaska R. Civ. Proc. 100 (2007); California Code §§ 3170, 3181, and 3182(2007); Fla. Stat. § 44.102(2007); Minn. Stat. Ann. § 518.1751 (2007) and Court Rule 114.04(a); N.H. Fam. Div. 2.13 (2007).

88 See www.marylandadrresearch.org/landscape/part1/court-of-special-appeals-1.

89 Md. Rule 8-206(c)(2015).

90 In 2011, the COSA ADR Division identified all US appellate mediation programs and determined that 56 appellate ADR programs existed representing 31 states and the US Virgin Islands. The COSA ADR Division received responses from 46 of the

56 identified and data from the courts with operational programs (40 out of the 46 respondents) formed the basis for the analysis. *An Analysis of the Maryland Court of Special Appeals ADR Division January 2012 Appellate Mediation Program National Questionnaire*, 3–4 (September 11, 2012).

91 *Id.* at 3–4.

92 Maryland's appellate mediation program, which conducts over 100 mediations per year, had an above-average resolution rate of 57.55%. *Id.* at 5.

93 Maryland reported the highest number of mediations in a year (187). Maryland's appellate mediation program, which conducts over 100 mediations per year, had an above-average resolution rate of 57.55%. *Id.* at 5–6.

94 See generally, Katz, Lucy V., *Compulsory Alternative Dispute Resolution and Voluntarism: Two-Headed Monster or Two Sides of the Coin?*, U. of Missouri School of Law J. of Dispute Res., Issue 1, Article 4 (1993).

95 *Id.* at 2.

96 *Short Bros. Const., Inc. v. Korte & Luitjohan Contractors, Inc.*, 356 Ill. App. 3d 958, 293 Ill. Dec. 444, 828 N.E.2d 754 (5th Dist. 2005).

97 See Richard D. English, *Alternative Dispute Resolution: sanctions for failure to participate in good faith in, or comply with agreement made in, mediation*, 43 A.L.R. 5th 545 (originally published 1996); see also *In re Air Crash Disaster at Stapleton Int'l Airport*, 720 F.Supp. 1433 (D.Colo.1988); *Abney v. Patten*, 696 F.Supp. 567 (W.D. Okl. 1987); *Official Airline Guides, Inc. v. Goss*, 6 F.3d 1385 (9th Cir.1993)(sanctions imposed on party for failure to have executive attend settlement conference); *Ferrero v. Henderson*, 2003 WL 21796381 (S.D. Ohio 2003)(F.R.C.P. 16 provides support for sanctions for failing to mediate in good faith); see also *Nick v. Morgan's Foods, Inc.*, 270 F.3d 590 (8th Cir. 2001) (imposing sanctions where party did not prepare required mediation memorandum and failed to have a representative with settlement authority at the conference).

98 *Strandell v. Jackson County*, 838 F.2d 884 (7th Cir.1987). Shortly after the *Strandell* decision, however, four other federal district courts ruled that judges *do* have authority to order parties to participate See *Home Owners Funding Corp. of Am. v. Century Bank*, 695 F. Supp. 1343 (D. Mass. 1988); *Federal Reserve Bank v. Carey-Canada, Inc.*, 123 F.R.D. 603 (D. Minn. 1988); *McKay v. Ashland Oil, Inc.*, 120 F.R.D. 43 (E.D. Ky. 1988); *Arabian Am. Oil Co. v. Scarfone*, 119 F.R.D. 488, 506 (M.D. Fla. 1988).

99 Federal courts with court-connected arbitration programs with specified monetary figures include the E.D. of Pennsylvania, D. of New Jersey, M.D. of Florida, W.D. of Oklahoma, E.D. of Washington, W.D. of Washington, and E.D. of New York. Except for the W.D. of Oklahoma, where the limit is $100,000, most of these federal district courts permit or mandate court-connected arbitration when damages do not exceed $150,000. States providing for mandatory court-connected arbitration include California, Oregon, New York, Washington, and Arizona.

100 See generally Richard D. English, *Alternative Dispute Resolution: sanctions for failure to participate in good faith in, or comply with agreement made in, mediation*, 43 A.L.R. 5th 545 (originally published 1996).

101 *Id.* (internal citations omitted).

102 *Foxgate Homeowners' Ass'n, Inc. v. Bramalea California, Inc.*, 26 Cal. 4th 1, 108 Cal. Rptr. 2d 642, 25 P.3d 1117 (2001).

103 See *Pallante v Paine Webber, Jackson and Curtis*, 1985 WL 1360 (S.D. N.Y. 1985) (granting motion to dismiss litigation, confirming arbitral award, and awarding defendant attorneys' fees and costs).

104 See *Shafii v British Airways*, 799 F Supp 292 (E.D. NY 1992)(stating court lacked jurisdiction due to the mandatory nature of the Railway Labor Act's arbitration procedures, which provide the 'exclusive forum' for dispute resolution).

105 See Richard D. English, *Alternative Dispute Resolution: sanctions for failure to participate in good faith in, or comply with agreement made in, mediation*, 43 A.L.R. 5th 545 (originally published 1996)(internal citations omitted); see also *Triad Mack Sales and Service, Inc. v. Clement Bros. Co.*, 113 N.C. App. 405, 438 S.E.2d 485 (1994) (entering default judgement against a defendant in a contract dispute action because only its attorney appeared at a court-ordered mediation).

106 Gutterman, Alan S., *Interaction of contractual ADR clauses with court-mandated ADR*, 2 Cal. Transactions Forms--Bus. Transactions § 14:5 (September 2015).

107 Senft, Louise Phipps, and Cynthia A. Savage, *ADR in the Courts: Progress, Problems and Possibilities*, 108 Penn. St. L. Rev. 327 (Summer 2003).

108 *Id*. at 336.

109 *Id*.

110 *Id*. at 340–348.

111 McAdoo, Bobbi, *A Mediation Tune Up for the State Court Appellate Machine*, Journal of Dispute Resolution, 333 (2010).

112 7 Bruner & O'Connor Construction Law § 21:3 (referencing Watson, Jr., *The Case for Mediated Case Management*, 1 Am. J. of Mediation 1 (2007).

113 McAdoo, *supra* note 44, at 333–334.

References

Abramowitz, Ava J. (2014) The Keys to Keeping a Project on Track, *Journal of the American College of Construction Lawyers*, 8(2), 43–58..

Black's Law Dictionary (2014) Tenth edition. New York: Thomson Reuters.

Bruner, Philip L. and O'Connor, Jr., Patrick J. (2015) *Bruner and O'Connor on Construction Law*. New York: Thomson Reuters.

Charkoudian, Lorig et al. (2016) *Impact of Alternative Dispute Resolution on Responsibility, Empowerment, Resolution, and Satisfaction with the Judiciary: Comparison of Short- and Long-Term Outcomes in District Court Civil Cases*, Annapolis, MD: Community Mediation Maryland, State Justice Institute.

Dettman, K.L., Harty, M.J. and Lewin, J. (2010) Resolving megaproject claims: Lessons from Boston's 'Big Dig', *Construction Law*, 30(2) 5–17.

English, Richard D. (1996) Alternative Dispute Resolution: sanctions for failure to participate in good faith in, or comply with agreement made in, mediation, 43 *A.L.R.* 5th 545.

Groton, James (1997) The progressive or 'stepped' approach to ADR: Designing systems to prevent, control, and resolve disputes. In Gaitskell, R (ed) *Construction Dispute Resolution Handbook*. London: ICE Publishing.

Gutterman, Alan S. (2015) Interaction of contractual ADR clauses with court-mandated ADR *California Transactions Forms—Business Transactions* 14(5).

Hafer, Randy, et al. (2010). *Dispute Review Boards and Other Standing Neutrals: Achieving 'Real Time' Resolution and Prevention of Disputes*, CPR Dispute Prevention Briefing: Construction. New York: International Institute for Conflict Prevention and Resolution (CPR).

Halls, Peter C. (2008) *General Conditions: A Comparison of Consensus DOCS and the Revised AIA Documents*, Minneapolis: Minnesota CLE.

Harris, Troy and Gavin, Donald (2013) *ADR in Construction United States of America*, London: IBA International Construction Projects Committee.

Ingleby, R. (1993) Court sponsored mediation: The case against mandatory participation, *The Modern Law Review Limited*, 56:3, 441–451.

Katz, Lucy V. (1993) Compulsory alternative dispute resolution and voluntarism: Two-headed monster or two sides of the coin? *University of Missouri School of Law Journal of Dispute Re*solution, 1993(1), Article 4. Available online at http://scholarship.law.missouri.edu/cgi/viewcontent.cgi?article=1022&context=jdr [Accessed 30 September 2016].

LaFountain, R., Schauffler, R., Strickland, S. and Holt, K. (2012) Examining the work of state courts: An Analysis of 2010 state court caseloads (National Center for State Courts 2012) Available online at: www.courtstatistics.org/Other-Pages/Examining-the-Work-of-State-Courts.aspx [Accessed 30 September 2016].

Legal Information Institute (n.d.) Alternative dispute resolution – state laws, Available online at www.law.cornell.edu/wex/table_alternative_dispute_resolution [Accessed 30 September 2016].

Malega, Ron, and Cohen, Thomas H. (2013) *State Court Organization, 2011*, Special Report of Bureau of Justice Statistics. Washington, DC: Us Department of Justice

Maryland Mediation and Conflict Resolution Office (2015) *Consumers' Guide to ADR Services in Maryland*. Annapolis, MD: MACRO.

McAdoo, Bobbi (2010) A mediation tune up for the state court appellate machine, *Journal of Dispute Resolution*, (2010)4, Article 5. Available online at http://scholarship.law.missouri.edu/cgi/viewcontent.cgi?article=1604&context=jdr   [Accessed 30 Septembert 2016].

McCallum, Mark H. (2008) Getting directly to the point of the contested matter: dispute mitigation & resolution in ConsensusDOCS Construction Forms, *A*merican Bar Association Forum on the Construction Industry, September 11–12.

National Center for State Courts (2015) *State Court Guide to Statistical Reporting*. Court Statistics Project, Version 2.1.1. Williamsburg, VA: National Center for State Courts.

Niemic, Robert J. (2006) *Mediation and Conference Programs in the Federal Courts of Appeals: A Sourcebook for Judges and Lawyers* (second edition). Washington, DC: Federal Judicial Center.

Reuben, Richard C. (2000) Constitutional gravity: A unitary theory of alternative dispute resolution and public civil justice, *UCLA Legal Review* 47(4): 949–1104.

Senft, Louise Phipps, and Savage, Cynthia A.(2003) ADR in the courts: Progress, problems and possibilities, *Pennsylvannia State Law Review*, 108, 327.

Stipanowich, Thomas J. (1998) The multi-door contract and other possibilities, Ohio State Journal on Dispute Resolution 13, 303–378.

Stienstra, Donna (2011) *ADR in the Federal District Courts: An Initial Report*. Available online at http://www.fjc.gov/public/pdf.nsf/lookup/adr2011.pdf/$file/adr2011.pdf [Accessed 30 September 2016].

Tobias, Carl, (1998) Did the Civil Justice Reform Act of 1990 actually expire?, *University of Michigan Journal of Law Reform* 31, 887

Tucker, Matthew Patrick (2005) Alternative dispute resolution use in the construction industry. Presented to the Faculty of The University of Texas at Austin.

United States Census Bureau. *Construction Spending*. Available online at www.census.gov/construction/c30/prpdf.html [Accessed 30 September 2016].

List of Cases

Brown v. Witco Corp., 340 F.3d 209 (5th Cir. 2003).

Ferrero v. Henderson, 2003 WL 21796381 (S.D. Ohio 2003)

In re Atlantic Pipe Corporation, 304 F.3d 135 (1st Cir. 2002)
Nick v. Morgan's Foods, Inc., 270 F.3d 590 (8th Cir. 2001)
Pallante v Paine Webber, Jackson and Curtis, 1985 WL 1360 (S.D. N.Y. 1985)
Shafii v British Airways, 799 F Supp 292 (E.D. NY 1992)
Strandell v. Jackson County, 838 F.2d 884 (7th Cir.1987)

List of Laws, Rules, Acts, and Statutes

Federal Arbitration Act, 9 USC. § 1, *et seq.*
Revised Uniform Arbitration Act
28 USC.A. §§ 651 to 658 (Judicial Improvements and Access to Justice Act).
28 USC. § 1491
United States District Court for the Northern District of California Local Rule 7
California Court Rule 5.83
Maryland Appellate Court Rule 8–206
Minnesota General Rule of Practice 114.04

6 Conclusions

A research roadmap for court-connected mediation

Deniz Artan Ilter

The key challenges to the widespread use of mediation in the construction industry have been investigated recently in a number of publications. Agapiou and Clark (2011) and Agapiou (2015) pointed to the apathetic and resistant attitude of lawyers towards mediation as a major barrier and suggested that a clearer definition of the role of the lawyer in mediation would help to overcome this perceived barrier. Ilter and Dikbas (2009), on the other hand, analysed the clients' experience with mediation and stated that lack of knowledge in the industry as well as sector based institutions and incentives appear as major barriers. To overcome these barriers, court-connected mediation has been initiated in some countries, especially in Anglo-American jurisdictions over the past 20 years, to encourage the parties to co-operate with each other before or during the trial and use mediation as the case allows. The motivation behind the introduction of court-connected mediation is reducing the number of cases reaching the court system. However, court connection deeply affects the nature of the process, and this situation is usually referred to as 'the inherent dilemma of court-connection mediation' in the literature. There have been very few studies on court-connected construction mediation and this publication is the result of a perceived need for further studies to examine the ways in which connection with formal civil justice systems influence mediation, the practice and performance of court-connected mediators and the ways lawyers approach mediation process across different legal systems and jurisdictions. This book investigates how court-connection have shaped mediation practice in different common-law jurisdictions encompassing England and Wales, South Africa, Hong Kong and the USA.

The aim of this chapter is to give an overview of the development of court-connected mediation in different jurisdictions, uncovering the commonalities as well as the differences in the experience and the approach in each specific country. Drawing together the main points of the provided national case studies, widening range of court-connected mediation possibilities available within the commercial disputes resolution sphere has been assessed and a research roadmap proposed for the further proliferation of the practice, focusing on the current barriers to the widespread use of court-connected mediation in the construction industry.

Evolution of court-connected mediation

In commercial disputes worldwide, as in others, the judiciaries struggle to keep pace with the mounting backlog of cases and allocate the considerable resources needed for remedy, while the public, the potential – and occasionally actual – litigants, are deeply concerned over the high costs and the often inconveniently slow processing associated with civil litigation. In many countries, the last few decades have seen a growing use of alternative dispute resolution methods, judiciaries and legislative bodies paving the way in a continuum from sanctioning and encouraging such methods to mandating them as a compulsory step within the litigation process, aiming to obtain a resolution before the commencement, or at any rate completion of litigation proper. Various types of mediation are among the most applied and sanctioned methods, considered by many to be particularly applicable to any conceivable type of dispute in the construction industry.

The Chapter on the USA describes how, starting with arbitration in the early-twentieth century, the USA has a long history of gradually formalizing and officially sanctioning processes that have later come to be subsumed under ADR. Only nine states currently lack dedicated dispute resolution statutes or rules. Construction industry being a major driving force behind the inception of the US ADR, today's construction disputes are typically resolved prior to trial, either through informal dispute resolution processes like mediation, non-binding arbitration, or court-administered ADR programs. Popularity of mediation among other processes is largely ascribed to private and confidential nature of proceedings and to its non-binding nature.

Court-connected ADR has been available in the US from 1970s and it is stated that the great variety of ways in which it is incorporated into federal and state judicial and administrative systems prohibit even descriptive generalizations. That legal rules affecting the practice of mediation, for instance, can be found in more than 2,500 statutes, thus risking conflict and inconsistency, has led related agencies to formulate a Uniform Mediation Act in 2002 and its eventual enactment in 12 states.

Significantly, non-profit and for-profit organizations, in addition to statutory and judicial efforts, have been instrumental in promoting and establishing ADR in general, and mediation in the construction industry, in particular. They have developed uniform rules, streamlined procedures with clear protocols and deadlines, including industry-specific rules like 'Construction Industry Arbitration Rules and Mediation Procedures' and 'Rules for Expedited Arbitration of Construction Disputes,' and maintain lists of trained mediators and arbitrators.

Developed by relevant independent agencies, the National Standards for Court-Connected Mediation Programs (NSCCMP) to guide and inform courts apply to all court-connected programs, and cover issues including availability of mediation services and educating parties and their attorneys about the program. Generally, the court will order the parties, often by way of a case management order, to complete, by a date certain, mediation or other court-connected ADR procedure, most often an attempt at pre-trial settlement involving an evaluation by a neutral expert. ADR, mostly in the form of mandatory mediation or settlement conferences has become widespread in various level federal courts as well, particularly after enactment of laws addressing the issue started in the 1990s, the Alternative Dispute Resolution Act of 1998, in particular. The chapter provides ample information, supported by statistics, on the court-connected mediation and other ADR in the federal courts.

The chapter on the situation in the UK and Wales regarding mediation expresses that these countries have a shorter history of mediation development compared to Australia and the USA. Another striking difference from particularly the US experience is the relatively much greater uniformity throughout the jurisdiction with respect both to statute and practice of mediation. From some voluntary action in the 1970s and 1980s, to the 1990s when the judges, with a view to abating court congestion, were recommended to issue ADR orders prior to trial, mediation has been connected through 1998 Civil Procedure Rules to litigation, indeed embedded into its structure. The chapter makes it clear that expansion in mediation practice has been achieved by connecting the costs rules, which empower judges to penalise litigants, the eventual winner included, if they unreasonably refuse to mediate in appropriate cases when one side has proposed its use. As a future perspective, the chapter draws attention to strong indications that mediation may soon become part of normal pre-trial phase by statute, no more requiring litigants' initiative or a court order as a case management gesture at the judge's discretion.

In the construction industry in particular, a 2014 amended Pre-Action Protocol for Construction and Engineering Disputes stipulates and formalizes how the parties should come together and seek settlement through ADR, and, if unsuccessful, how the judge may invite both to reconsider ADR, most often mediation, negotiation or neutral evaluation by a judge. Upon request of the parties, the judge can provide a non-binding evaluation of the case, likely outcome at trial, and opinion on appropriate settlement. Somewhat similar to the US practice, courts mostly do not differentiate between alternative procedures, which has blurred the edges between settlement negotiation, mediation and dispute resolution involving evaluation.

The chapter also points to the emergence of discussions of principle regarding merits or else of litigants being 'coerced', under the threat of costs penalties, into mediation or other ADR by court orders, arguably in defiance of the core values of mediation and to how in some cases mediation can be quite costly as well considering high fees required by qualified mediators. These points notwithstanding, it is understood that mediation, in general, is deemed a viable and rather successful method, significantly reducing the burden of the courts and apparently serving the construction industry, 72 per cent of whose lawyers, for instance, declared satisfaction in a survey with the method. Other research shows, however, that nearly half of lawyers say they never or rarely initiate mediation unless compelled by the court. Further, it is said, some lawyers use the process in a more legalistic and adversarial way by presenting their clients' cases and using legal arguments to strengthen claims.

The chapter on South African practice describes how mediation was introduced to the South African construction industry in 1976 and how efforts to transform it into court-connected mediation started in 2003 leading to enactment of pertinent legislation in the last decade. The Civil Justice Reform Project of 2012, aiming to simplify court processes and implement mandatory mediation to settle cases out of court, is familiarly driven by an effort to alleviate court congestion. Court-connected mediation in a limited number of pilot courts finally began in 2014, wielding punitive costs to persuade litigants to settle out of court. It remains to be seen whether such a move will spike utilization of mediation, which until that time has been rather sparse.

It is emphatically noted that the ADR practice in the UK and its construction industry in particular, has served as a model and example in South Africa: somewhat similar, perhaps, to the UK following the US example in some respects. Attention, however, is drawn to a peculiarity of mediation in the South African construction industry in that the mediators are normally expected to assume an evaluative rather than a facilitative approach and to carry out a conciliatory rather than a mediating function. It should be stated that such a tendency is also significantly noted (if not with the same degree of emphasis) in the chapters on UK and the US. Considerably established informal practice in the private construction companies involving a neutral professional, predominantly architectural professionals, conciliating disputes may reasonably be expected to persist despite the changing legal landscape. Arbitration, on the other hand, has been a traditional backbone in ADR practice of the sector, a non-binding type of adjudication being yet another fairly popular mechanism. Again, a non-binding variant of arbitration is another quasi-mediation activity with some popularity. A real challenge, however, is posed by the public sector, which supports the private industry with excessive contracts: when party to any contract the public representatives are known to scrupulously replace any reference to ADR with litigation.

For disputes in the construction industry, the latest practical situation seems to be as follows, according to the Joint Building Contracts Committee Principal Building Agreement (JBCC PBA) of 2014: if direct negotiations or conciliation fail, parties will be required to submit to mediation or any other form of ADR offered in the contract. Should they fail to resolve the dispute, they would submit to arbitration. Without a JBCC PBA contract, parties would be required to submit to court-connected mediation. Qualifications and costs of mediators in court-connected mediation and fear of political influence and corruption seem to be serious points of concern and as such they constitute part of pitfalls and uncertainties that await court-connected mediation in South Africa.

The chapter on Hong Kong also reports on the situation in Singapore and many jurisdictions in Australia as well, where the practice of mandatory court-connected mediation concurrently with voluntary mediation has generally obtained positive results in reaching favourable settlement rates. Mediation is a viable ADR process in Hong Kong, too. The Mediation Ordinance enacted in 2012 and aiming to promote mediation and secure confidentiality of mediation communication has reinforced the position of mediation as a viable and popular form of ADR.

Mediation is also the most commonly used form of ADR in construction dispute resolution. Form contracts, issued by various agencies and routinely used by the construction industry in Hong Kong, commonly specify mediation within a staged ADR scheme. Furthermore, under Practice Decision 6.1 (PD 6.1) of the Hong Kong judiciary, an 'indirect' court-connected mediation is in force for cases falling under the 'Construction and Arbitration List', where the judge has the authority to deter, through costs orders, those who unreasonably abstain from even minimal participation in mediation.

PD 6.1 appears to be a special case, however, since after extensive fact-finding studies, pilot projects and political debate a resolution has been reached in Hong Kong, opting against adoption of a general mandatory court-connected mediation. A governmental report of 2010 stated that it would be desirable to wait for the impact of the Civil Justice Reforms of 2009 (CJR) on the use of mediation in Hong Kong to be analysed and that the question of compulsory mediation be revisited in the future rather than introducing without a consensus view of the stakeholders. This decision has particular significance in the light of various queries and criticism directed against mandatory court-connected mediation by some members of the industry, by law professionals and others in all countries reported in the chapters.

Some of the objections raised against mandatory court-connected mediation pertain to issues of practical utility such as that it risks reducing settlement rates or expected gains in time and costs. Some however, relate to deeper lying, more fundamental points and criticize emphasizing utility instead of rule of law and fundamental values reflected in the rule of law, such as equality and access to justice under the law. That vein of criticism, in fact, would apply to ADR not only in its mandatory form.

A research roadmap for court-connected mediation

Evolution and current practice of court-connected mediation in the pioneer jurisdictions have been analysed throughout the book. Both commonalities and differences in the experience and approaches of court-connected mediation provoke new questions for the future of this method in the countries analysed as well as others that will follow. This section outlines the main issues tackled in the national case studies and presents a research roadmap highlighting the areas requiring attention for a more widespread use of court-connected mediation.

One such area is concerned with whether the use of court-connected mediation is increasing the uptake of voluntary mediation. Brooker points out the probability that court-connection may be diminishing many of the core features and benefits of mediation such as self-determination, speed, cost savings and creative outcomes which usually is impaired by the involvement of attorneys in the process. Stipanowich states that the mandatory nature of court-connected programs has attracted criticism because it seemingly 'forces' litigants into a non-traditional process that they have not chosen. Cheung also draws attention to the concern that the benefits of voluntary mediation, such as high settlement rates, cannot be maintained with mandatory mediation. It is clear that without promising increases in settlement rates, court-connected mediation can only lead to higher costs and more waiting time at the end because the disputants will need to bear both the mediation and litigation processes. Therefore, a core area of future research in the court-connected mediation domain is the investigation and quantification of the advantages and benefits claimed for court-connected mediation.

Another area that requires attention is the contractual use of mediation. In Hong Kong, the use of mediation within the court system has been effected through the promulgation and application of cost orders for the parties who unreasonably refuse or fails to mediate to the minimum participation level when there is a mediation notice. However, it is not yet clear what outcomes parties should expect from a mediation clause, as what makes 'unreasonable refusal' of mediation seems to differ from court to court. A review of the perceptions on contractual mediation and its importance within the context of court connection is necessary, as this may be the way to expect proliferation of court-connected mediation in many countries such as Hong Kong.

Cases where mediation, including court-connected, is not suitable constitute a third area that deserves attention. The cases provided throughout the chapters indicate that the range of disputes suited to court-connected mediation is quite wide. Some researchers argue that mediation is suitable for all commercial disputes and this unboundedness provides a convenient setting for a possible future scenario where mediation is prescribed as a formal initial stage in all commercial disputes. However, Cheung draws attention to the right of the disputants to having their disputes decided in courts. Many common-law countries have implemented court rules to encourage or use mediation and penalties for non-use. Some countries and specialized construction courts have their own mediation schemes or programmes to promote, govern and monitor activity. However, a closer look and a further classification of the type of disputes in which a court-connected mediation should be 'encouraged' or 'punished for non-use', could lead to higher success rates in these schemes and programs.

All contributors state that the style of mediation deeply affects the outcomes. Du Preez argues that mediators have concentrated on the evaluative process with limited regard for the soft skills relating to mediation in South Africa. This results in addressing the technical issues without feeling the need to consider the psychological component of mediation. Therefore, another area of concern in this context is the mediator styles employed in court-connected mediation, which may tend to differ from voluntary mediation, and the effects this may have on the success rates of the process.

And finally, research into the various stakeholders' perceptions on court-connected mediation is required. This has been a popular topic in the voluntary mediation domain. The bulk of academic research seeking to explore the views and experiences of key actors involved in mediation has usually been concerned with the legal professionals rather than clients or other players in the dispute resolution game. As Agapiou suggests, the reason behind this tendency might perhaps be related to the fact that lawyers are an easier grouping for researchers and promotional bodies to access than the client base. As lawyers are seen as the gatekeepers to the practice of mediation, their perceptions are vital in development of the process. To what extent and when lawyers are in the driving seat relative to decisions in court-connected mediation is an altogether more nuanced issue, however, that warrants further analysis and international comparison.

Reduced time, cost, and confidentiality have always been core features of mediation. While court-connected mediation in many jurisdictions is still in its infancy, as Stipanowich points out, proponents would argue that the introduction of court-connected mediation added 'judicial economy' to that list of benefits. Although case backlogs may be the primary impetus for court-connected mediation, in the US experience, the notion of self-determination has also contributed to its rapid growth and success. Stipanowich also draws attention to the possibility that mediation in litigation and arbitration can be 'an effective case management process by which to reach consensus of the parties on deposition discovery, scheduling, hearing time allotted to each party, and how the case otherwise might most efficiently be tried.'

As construction practitioners and lawyers working in the area grow more aware of the benefits of court-connected mediation, more disputants that may have been reluctant to submit to mediation will be willing to participate in settlement of cases that would otherwise have remained embroiled in arbitration or litigation. As court-connected mediation programs develop in more jurisdictions, and the progress is supported by scientific approach and research, it is likely the ways to maximise the benefits of court-connected mediation will crystallize.

The authors hope that the overview of the evolution of court-connected mediation in different jurisdictions and the research roadmap proposed to tackle the current barriers identified in this book provide some significant contribution to the proliferation of the area.

Index

Italic page numbers indicate tables; **bold** indicates figures.